Fractional Dynamics, Anomalous Transport and Plasma Science

Christos H. Skiadas
Editor

Fractional Dynamics, Anomalous Transport and Plasma Science

Lectures from CHAOS2017

Springer

Editor
Christos H. Skiadas
ManLab, Technical University of Crete
Chania, Crete, Greece

ISBN 978-3-030-04482-4 ISBN 978-3-030-04483-1 (eBook)
https://doi.org/10.1007/978-3-030-04483-1

Library of Congress Control Number: 2018961726

This Springer imprint is published by the registered company Springer Nature Switzerland AG
The registered company address is: Gewerbestrasse 11, 6330 Cham, Switzerland

Preface

Organizing this book on *Fractional Dynamics, Anomalous Transport and Plasma Science—Lectures from CHAOS2017*, we have to face the fascinating theory and applications of plasma science. The development of the related theory was followed by numerous applications, whereas the new findings brought new questions and new scientific areas of research and model building. It was evident that the classical model building based on derivatives had to be improved by using a "fraction" of some of these derivatives. Though the expansion of a function in a Taylor series theoretically could include as many as higher-order derivatives as possible to account for the details of the system, the usual methodology includes the first- and second-order derivatives in the majority of cases. The introduction of stochastic differentials and stochastic calculus added extra terms to compensate for the stochastic character of the real-life situations. Following the original works after 1900 and the first interesting achievements in stochastic model building the precise data collection give rise to findings needing a model reorganization and improvement. Anomalous transport was the case of many findings where the theory should be improved. This was also supported by the development of chaos and nonlinear science. The need for an advanced model building brought into light an almost forgotten mathematical theory related to fractional calculus. Fractional dynamics are extensively developed in the last 30 years with numerous publications along with advancements of fractional calculus.

However, more work should be done. A part of the expected improvements has to do with the advancements in fields like the turbulent theory, the statistics far from equilibrium and perhaps the expected advancements with the plasma research in cosmology. Even more the fractional dynamics methodology should be expanded to include not only the fractional derivatives theory as it is already developed but also new achievements. The papers selected for this book further from including the plasma history and fractional dynamics and anomalous transport in micro- and macroscale provide material for a better understanding of the underlying theory and practice.

Referring to the definitions of plasma in nowadays, the main part considers plasma as the fourth state of matter of an ionized form under "hot" or "cold" temperature conditions. Hot refers to temperatures like those inside the sun, a classical plasma formation, whereas cold plasma refers to low temperature of heavy species (ions and neutrals) and of high-temperature electrons. The latter case can lead to a plasma application in room temperature as is the fluorescent lamp and other plasma applications.

With fractional dynamics, we consider dynamical systems governed by fractional-differential equations containing derivatives of non-integer order. Fractal properties are also considered. Anomalous transport mainly refers to turbulence, complex particle orbits, fluctuations, transport barriers, locality, super-diffusion and sub-diffusion. The study could be accomplished by solving a kinetic equation or a stochastic Fokker–Planck equation in a classical or in a fractional form.

The Origins of Plasma Science in connection to electricity and electric charges and devices leading to arc plasma are explored in the first chapter of the book from Jean-Marc Ginoux and Thomas Cuff. The historical review starts from around 585 BC when Thales of Miletus observed discharges due to frictional charge up. Discharges by frictional electricity were rediscovered and greatly expanded in the seventeenth and eighteenth century by many scientists. It follows the capacitor invention or the "Leyden experiment". The problem to increase the size of the battery to reduce their internal resistance so that to move from a simple spark to an arc and a **continuous arc "PLASMA"** achieved by Vasily Petrov, 1803, and then by Sir Humphry Davy, 1808. Thus, the first electric street lighting employing carbon arc lamps, also called "electric candle", was developed in 1875. In 1863, James Clerk Maxwell, by mathematical reasoning, formulated the theory that light and radiant heat were electromagnetic phenomena, caused by strains set up in the all-pervading medium, similar to the electric lines and magnetic lines theory proved by Heinrich Hertz, 1887.

The second important connection with plasma was magnetism and the most widely used system called as **Magnetron**. Victor J. Law and Denis P. Dowling in the second chapter review the magnetron electronic valve family development including a large variety of devices. They explain why and how the magnetron has its origins in the conflict around the patents of thermionic diode valve in the early 1900 when to prevent potential litigation problems, Governments and industrial companies around the world began research programs to replace the convoluted problem of grid electrode with a magnetic field to control current flow with the aim to produce space-charge oscillations travelling through the flux density of the magnetic field.

Christos H. Skiadas and Charilaos Skiadas presented and applied the diffusion theory as was developed based on a relatively difficult and laborious methodology. However, the advances in mathematics and the use of computers in applications gave rise to interesting applications in various scientific fields. The main lines of the theory include a stochastic model and stochastic paths. The main task is to find the mean value and the variance σ. Then, the Fokker–Planck Equation should be solved with appropriate boundary conditions to find the transition probability density and then the first exit time or hitting time probability density. The classical Fokker–Planck (FP) equation has the form:

$$\frac{\partial p(h_t,t)}{\partial t} = -h_t \frac{\partial p(h_t,t)}{\partial h_t} + \frac{\sigma_t^2}{2} \frac{\partial^2 p(h_t,t)}{\partial h_t^2}$$

Two main fractional forms arise: The fractional FFP by replacing the second-order differential with the fractional one expressed by a fractional parameter α,

$$\frac{\partial p(h_t,t)}{\partial t} = -h_t \frac{\partial p(h_t,t)}{\partial h_t} + \frac{\sigma_t^2}{2} \frac{\partial^a p(h_t,t)}{\partial h_t^a}$$

and a fractional FFP by replacing the time derivative with the fractional one expressed by γ.

$$\frac{\partial p^{\gamma}(h_t,t)}{\partial t^{\gamma}} = -h_t \frac{\partial p(h_t,t)}{\partial h_t} + \frac{\sigma_t^2}{2} \frac{\partial^2 p(h_t,t)}{\partial h_t^2}$$

In both cases of the FFP equations, the solution methodology differs. The FFP may differ considerably when applying in various cases or various scientific fields, but the solution methodologies are similar usually leading to various forms including in the majority the gamma function. However, we have presented an example where the fractional theory could apply after solving a classical Fokker–Planck equation and afterwards to select a fraction of the derivatives. The case refers to a first exit time or hitting time probability density function related to a fraction of first- and second-order derivatives, the latter contributing with a fraction k to the stochastic process.

Anomalous diffusion by the fractional Fokker–Planck equation and Lévy stable processes are studied by Johan Anderson and Sara Moradi. They consider the motion of charged particles in a three-dimensional magnetic field in a cylindrical domain in the presence of linear friction modelling collisional Coulomb drag and a stochastic electric field according to the Langevin equations. This work is a review of current developments in modelling anomalous diffusion using a Fokker–Planck description with fractional velocity derivatives and Langevin dynamics where Lévy fluctuations are introduced to model the effect of nonlocal transport due to fractional diffusion in velocity space. Distribution functions are found using numerical means for varying degree of fractionality of the stable Lévy distribution as solutions to the Fokker–Planck equation and are compared to results from Langevin simulations. The statistical properties of the distribution functions are assessed by a generalized normalized expectation measure and entropy in terms of Tsallis statistical mechanics.

Analysis of low-frequency instabilities in low-temperature magnetized plasma is presented by Dan-Gheorghe Dimitriu, Maricel Agop. The authors refer to experimental results in the Q-machine at the University of Innsbruck in Austria while they provide analytical formulation of the related theory. The theoretical calculations follow fractal curves. They assume that the Q-machine magnetized plasma particles are moving on continuous and non-differentiable curves (fractal curves). A theoretical model was developed in the frame of the scale relativity theory. The model is able to explain some characteristics of the potential relaxation instability and the electrostatic ion-cyclotron instability, as well as the interaction between these two instabilities which leads to the amplitude and frequency modulation of the second instability by the first one. Experimental results are shown, which are in agreement with the theoretical model predictions.

Stefan Irimiciuc, Dan-Gheorghe Dimitriu, Maricel Agop propose a theoretical model attempting to explain the dynamics of charged particles in a plasma discharge where there is a strong flux of electrons from one plasma structure to another. Basically, the dynamics of the electrons are described using a forced damped oscillating system, with the aim to investigate the response of the global discharge current to different changes in resolution scale, oscillation frequency and damping coefficient. Based on a non-differential approach, an explanation was proposed for the modulated oscillation of plasma structure created in a spherical cathode with an orifice. Within the framework of the model, the particles move on fractal curves which might lead to an oscillating state of the charged particles. The system evolves from a double-period state towards a chaotic signal but never reaching it. The evolution of the system is "controlled" by the damping of the system and by the maximum frequency reached during a particular simulated time-series.

The theory and applications of fractional derivatives in many-particle disordered large systems are explored and applied by Z. Z. Alisultanov, A. M. Agalarov, A. A. Potapov, G. B. Ragimkhanov. In this chapter, they consider several possibilities for introducing fractional derivatives with respect to time and spatial coordinate into the equations of a many-particle system. They introduce a fractional time derivative in the quantum-mechanical Heisenberg equation, as well as elucidating the physical conditions for the appearance of fractional derivatives with respect to the spatial coordinate in the equation for the quantum Green's functions. For the latter, the Hartree–Fock approximation is used to calculate the interparticle interaction potential. Finally, the fractional derivative approach was applied for specific tasks in plasma science. Using an approach based on a fractional-order kinetic equation on the time variable, they investigated two types of instability in a gas discharge: the instability of the electron avalanche and the sticking instability in a non-self-sustaining discharge.

They further emphasize:

The appearance of fractional operators in the equation for the Green's function is related with the non-ideality of the system. The consideration was carried in the general form. It can be applied to both fermionic and bosonic systems. The most interesting is the application of a fractional-differential approach to a phonon gas. As is known, the non-ideality of the phonon gas, which is related to the complexity of the crystal structure, leads to anharmonicity, with which many interesting effects are related. The theory of anharmonicity of a fractional character has some peculiarities and can be of great interest in describing real experimental results.

For example, the elementary calculation shows that the fractional-differential approach gives a wide class of temperature dependence of heat capacity. At the same time, it is known that the temperature dependence of the specific heat for complex crystals is not a single-valued function, but is determined by the type of crystal and the structure. The latter circumstance is practically not described by the existing theory. Thus, the fractional-differential approach in the theory of anharmonic effects in crystals is of great interest.

And conclude that:

Note that most physical models (classical and quantum) with fractional derivatives are currently at the stage of intensive development. Problems arise even at the stage of choosing one or another fractional operator. In general, preference is given to those models that most adequately describe the available experimental data. The specificity associated with fractional derivatives can manifest itself in the stochastic dynamics and kinetics of large systems. It is logical to assume that simultaneous introduction of fractional derivatives with respect to time and coordinate in the classical and quantum cases is required. However, these issues require further development!

Maricel Agop, Alina Gavrilut and Gabriel Crumpei consider and explore motions of the physical systems that take place on continuous but non-differentiable curves (fractal curves). Since the non-differentiability becomes a fundamental property of the motions' space, a correspondence between the interaction processes and multifractality of the motion trajectories can be established. Then, for all scale resolutions, the geodesics equations (in the form of the Schrödinger equation of fractal type) and some applications (similarities between dynamics at atomic and cosmic scales) are obtained.

Finally in the last chapter, S. L. Cherkas and V. L. Kalashnikov consider the perturbations of plasma consisting of photons, baryons and electrons in a linearly expanding (Milne-like) universe with taking into account the metric tensor and vacuum perturbations. Here, they use the oversimplified model of plasma as a pure radiation. They expose the scenarios of primordial baryon–photon plasma evolution within the framework of the Milne-like universe models. Such models find a second wind and promise an inflation-free solution of a lot of cosmological puzzles including the cosmological constant one.

I would like to thank the authors and contributors of this volume, Christian Caron from Springer and the other staff of Springer for a continuing collaboration started several years ago leading to important publications.

Athens, Greece
October 2018

Christos H. Skiadas

Contents

From Branly Coherer to Chua Memristor

Jean-Marc Ginoux and Thomas Cuff

1 The Origin of Arc Plasma Science

According to Anders [1] "Since there have been two distinct developments of electrical energy sources – the capacitor and the electrochemical battery – the distinction between pulsed and oscillating and continuous arc discharges appeared quite natural."

1.1 Pulsed and Oscillating Arc Discharges

Around 585 BC, Thales of Miletus discovered that if he rubbed amber (called elektron in Ancient Greek) with a piece of fur, that amber could attract lightweight objects (like feathers) to itself. He was thus one of the first to observe discharges due to frictional chargeup. Subsequently, discharges by frictional electricity were rediscovered and greatly expanded in the 17th and 18th century by many scientists such the famous Otto von Guericke (1602–1686), Francis Hawsksbee (1666–1713) and Samuel Wall to name but a few. Von Guericke is credited with inventing before 1663 a primitive form of frictional electrical machine consisting of a sulphur globe attached to an iron rod that he called "Elektrisiermaschine". By rubbing a sulfur globe with a dry hand,

J.-M. Ginoux (✉)
Université de Toulon (TLN), La Garde, France
e-mail: jmginoux@orange.fr

J.-M. Ginoux
Laboratoire d'Informatique et des Systmes, UMR, CNRS 7020, Marseille, France

J.-M. Ginoux
Archives Henri Poincaré, UMR CNRS 7117, Nancy, France

T. Cuff
Frederick Community College, Frederick, MD 21701, USA

C. H. Skiadas (ed.), *Fractional Dynamics, Anomalous Transport and Plasma Science*, https://doi.org/10.1007/978-3-030-04483-1_1

his charged globe was then able to attract and repel after contact other light objects such as feathers, drops of water and also to generate sparks. A few years after, the French astronomer, Jean Picard (1620–1682) made an amazing discovery:

> Towards the year 1676, Monsieur Picard was transporting his barometer from the Observatory to Port Saint Michel during the night, [when] he noticed a light in a part of the tube where the mercury was moving; this phenomenon having surprised him, he immediately reported it to the sçavans,[1] ...

After learning of the phenomenon from Johann Bernoulli (1667–1748), the Englishman Francis Hawksbee investigated the subject extensively. This led him to the discovery of the phenomenon "barometric light" that he called "body of fire" which revealed the possibility of electric lighting. We will see in the next section the importance of electric lighting in the development of *Wireless Telegraphy*. However, it was only half a century later that Benjamin Franklin (1706–1790) proved, with his famous kite, that a lightning strike was nothing else but an electrostatic discharge. In the meanwhile, a London physicist named Samuel Watt observed that while rubbing with his hand in the dark the head of his cane made of amber, it became luminous. Thus, he also suspected a connection between frictional electricity and lightning.

It was during this period that the capacitor was invented. There are many various versions about the history of this invention. At first, historiography credited Pieter van Musschenbroek (1692–1761), professor at University of Leiden (also spelled Leyden), for this invention that he described in a letter sent on January 20, 1746 to René Antoine Ferchault de Réaumur (1683–1757), his correspondent at the Royal Academy of Sciences in Paris. This letter was read by the French clergyman and physicist Jean-Antoine Nollet (1700–1770) who gave to Musschenbroek's experiment the name of "Leyden experiment". Nowadays, some historians consider that this invention was made originally by Ewald Jürgen von Kleist (1715–1759), Dean of the cathedral at Cammin in Pomerania (Germany) and so, there are no reasons to keep on calling it the "Leyden experiment". However, in June 1954, Pr. C. Dorsman found in the Philosophical Transactions of the Royal Society (London) a letter "from Mr. Trembley F.R.S. to Martin Folkes Esq; Pres. R.S. Concerning the Light caused by Quicksilver, shaken in a Glass-Tube, proceeding from Electricity". This letter is dated: Hague, 4 Feb. 1745 N.S. (New Style), but was not read in a meeting of the R.S. before Feb. 13 1746. In 1957, he published this letter with Crommelin [3] at the end of which one can read:

> There is an experiment that Mr l'Allamand has tried; he electrify'd a tin Tube, by means of a glass Globe; he then took in his left Hand a Glass full of water, in which was dipped the End of a Wire; the other End of this Wire touched the electrified tin Tube: He then touch'd, with a Finger of his right Hand, the electrified Tube, and drew a Spark from it, when at the same Instant he felt a most violent Shock, all over his Body. The pain was not always equally sharp, but he says, that the first time he lost the Use of his Breath for some Moments; and he then felt so intense a Pain all along his right Arm, that he at first apprehended ill Consequences from it; tho' it soon after went off without inconvenience. It it to be remarked, that in this Experiment he stood simply on the Floor, and not upon the Cakes of Resin.

[1] See Picard [2].

> Mr. Musschenbroek the Professor has repeated this experiment, holding in his Hand a hollow Bowl exceeding thin, full of water; and he says he experienced a most terrible Pain. He says, the Glass must not be at all wet on the outside.

Thus, they concluded that "as early as the very beginning of the year 1745 (that is nine months before von Kleist's experiment) the Leyden jar was invented at Leyden, by Allamand or by Musschenbroek or by both of them, exactly a year earlier than the well-known experiments" described by Musschenbroek in his letter to Réaumur dated from January 20, 1746 (Fig. 1).

Hence, electricity and discharge phenomena became a subject of science as well as amusement as exemplified by the famous anecdote according to which Nollet electrified two hundred person in the Hall of Mirrors (Galerie des Glaces) of Versailles before his majesty the king Louis XV. Then, at the end of the 18th the Italian physicist Alessandro Volta (1745–1827) gave to the Leyden jar the name of capacitor. In 1771, he had improved and popularized the electrophorus: a capacitive generator used to produce electrostatic charge via the process of electrostatic induction. Then, he published his research about the electrophorus in the *Philosophical Transactions of the Royal Society of London* in 1782 and they were reproduced in his Complete Works in 1816. In this paper Volta [4] wrote:

Fig. 1 The Leyden jar

> Questo artificio, voi forse già l'indovinate, consiste a riunire all eletrometro medesimo il *Condensatore*.[2]

Then, he exposed the relationship between the tension v, the charge q and the capacity C of the capacitor:

> Ciò che abbiam detto comprendersi facilmente che la tensione debbe essere in ragione inversa delle capacita, ci viene poi mostrato nella maniera pi chiara dall esperienza.[3]

In a previous publication we have suggested to call this relationship, which can be expressed in a modern formulation as $q = Cv$, "Volta's law". See Ginoux and Rossetto [5]. On March 20th 1800, Volta wrote his first letter to Joseph Banks (1743–1820) in which he described the first true battery which came to be known as the Voltaic Pile (see Fig. 2). This letter was sent for publication to the *Philosophical Transactions* and read at the *Royal Society of London* on June 26th 1800.

On November 7th and 20th 1801, Volta presented his device at the Institut de France before Napoléon who awarded him a gold medal. Although Volta made use of the term "tensione" in his articles introducing thus in a qualitative way the *voltage*, it was the French Physicist and Mathematician André Marie Ampère (1775–1836) who defined accurately the concepts of *current* and *voltage* in his first mémoir of 1820 [7] wrote:

> L'action électromotrice se manifeste par deux sortes d'effets que je crois devoir d'abord distinguer par une définition précise. J'appellerai le premier tension électrique, le second courant électrique.[4]

1.2 Continuous Arc Discharges

According to Anders [8] "Continuous discharges could only be obtained after enduring energy sources became available, namely in the form of a battery of electrochemical cells, invented by Volta in late 1799." Nevertheless, the very first Voltaic piles or batteries were not able to sustain continuous arc discharges due to their internal resistance. Then, a race of making large batteries started all over the world between the countries where the news of Volta's invention was known. This "sort of experimental fever" for the piles is exemplified by Ayrton [9] who wrote in her chapter entitled "A short history of the arc":

> Paper after Paper was written describing new and interesting results obtained with the pile. So numerous were these Papers in the course of the next year that in the middle of 1801 a certain Dr. Benzenberg wrote to the editor of Gilbert's Annalen:

[2] "This device, you probably already guessed, is to bring together all the same electrometers under the name *Capacitor*."

[3] "What we have said can easily be understood that the voltage ought to be in inverse proportion to the capacity, as it is shown clearly by experiments."

[4] "The electromotive action is characterized by two kinds of effects that I believe I must first distinguish by a precise definition. I will call the first voltage, the second electric current."

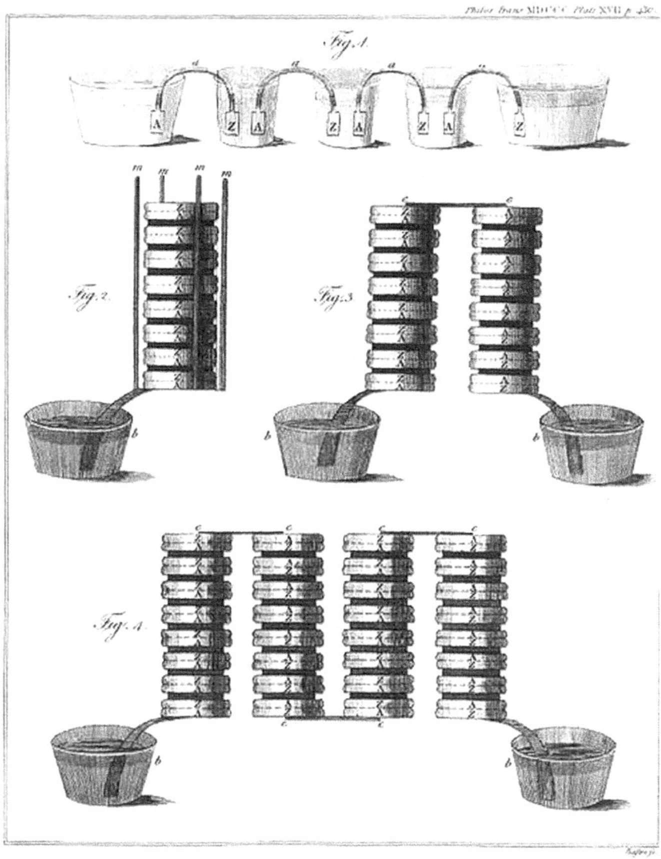

Fig. 2 Voltaic pile [6]

> Could not the Annalen, in consideration of its object, be a little more varied? Galvanism, interesting as it is, is still only a very small part of physics. We can apparently only expect any real advance in knowledge from such work as is carried out on a large scale, and not from each experimenter, whose slight knowledge and small apparatus allow him to discover only what ten others have already found out before him.

Among the various experiments performed with Voltaic piles (effect of the current on living things, decomposition of matter, …) those which dealt with the heating power of the current, more particularly with the sparks produced by making or breaking a circuit in which current was flowing led directly to the discovery of the continuous arc. As previously mentioned, the circumstances of this discovery were subject of much research. More particularly the questions of who discovered the continuous arc discharges and when they have been discovered are still debated. However, Humphry Davy (1778–1829) is generally credited with this discovery as recalled by Ayrton [9]:

> Later, in a lecture before the Royal Institution, given in 1801, Sir Humphry mentioned that the spark passing between two pieces of well-burned charcoal was larger than that passing between brass knobs, "and of a vivid whiteness; an evident combustion was produced, the charcoal remained red hot for some time after the contact, and threw off bright coruscations [10]."

Nevertheless, according to Ayrton [9] the internal resistance of the battery used by Davy "was very great compared with what it should be in order to maintain an arc and the passing of a spark would so lower the P.D. between the terminals that no other spark could pass till the battery had somewhat recovered." As a consequence, Davy's above description is not that of a continuous arc discharge but of a spark. At that time, the problem was to increase the size of the battery to reduce their internal resistance. However, it seems that it was solved by an unknown Russian scientist. In an article published thirty-five years ago, Kartsev [11] disclosed that Vasilii Valdimirovich Petrov (1761–1834) of St. Petersburg made experiments with carbon arcs in 1802 during which he observed continuous arc discharges. Then, he explained:

> Petrov made an original improvement which overcame the earlier insurmountable obstacle for the construction of such big piles - he placed the pile horizontally in three separate boxes. The horizontal position of the pile permitted to avoid pressing-out of the electrolyte from between the plates and removed limits for the pile's height and capacity. The pile constructed by Petrov was undoubtedly one of the biggest piles of that time [11].

Then, he published his results in 1803 in a book entitled "News of the galvani-voltaic experiments which professor of physics Vasily Petrov had conducted by means of a particularly huge battery consisting at times of 4200 copper and zinc disks and installed at St. Petersburg Medicine and Surgery Academy" (see Fig. 3). As pointed out by Kartsev this book "was by chance discovered in a library in the town of Vilno at the end of the ninetieth century".

Petrov wrote in his manuscript:

> If two or three charcoal pieces are placed on a glass plate or on a bench with glass legs, and if the charcoal is connected to both ends of an enormous battery using metallic but isolated conductors, and if the two pieces are brought in close distance of one to three lines [2.5–7.5 mm], then a very bright cloud of light or flame shines, burning the charcoal more or less fast, and one may illuminate a dark room as bright as one wants to [12].

Thus, according to Anders [8]: "Petrov had made, observed, and described [independently, and earlier than Davy] the first continuous arc discharge. Moreover, he suggested that the bright light or "flame" (plasma) could be used for lighting purposes, the first possible real application of electricity apart from entertainment of aristocrats."

About six years after Petrov's publication, and very likely unaware of it, Davy pursued his experiments. According to Ayrton [9], Sir Humphry Davy's laboratory manuscript notes for the years 1805–1812 were collected by Michael Faraday and then published in the Royal Institution. Concerning the years 1808 and 1809, one could read:

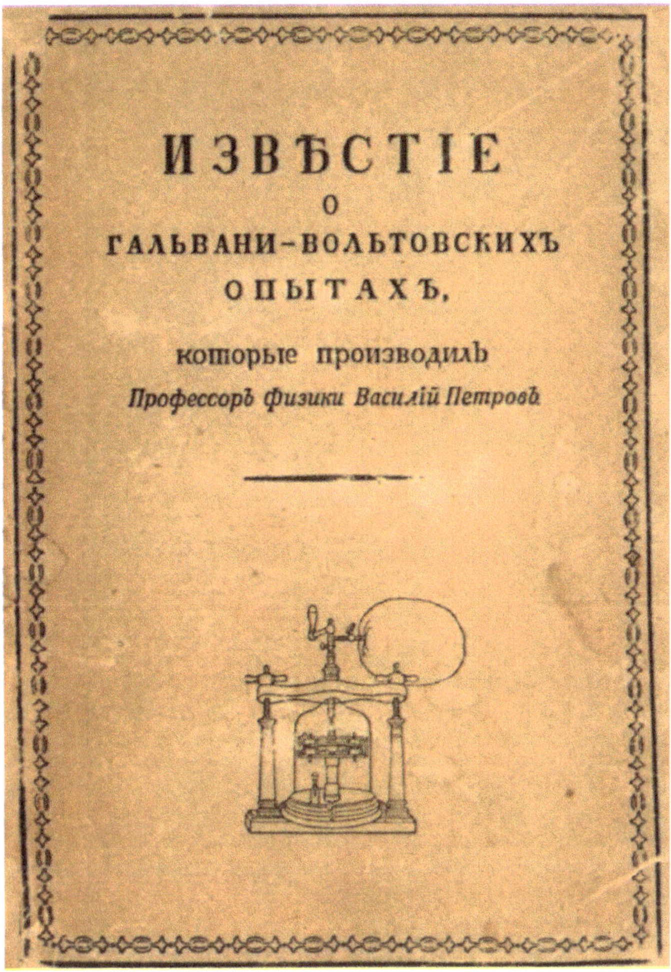

ИЗВѢСТІЕ
о
ГАЛЬВАНИ-ВОЛЬТОВСКИХЪ
ОПЫТАХЪ,
которые производилъ
Профессоръ физики Василій Петровъ

Fig. 3 Front page of V. V. Petrov's book [12]

April 20, 1808.

A given quantity of muriatic acid gas was acted upon by dry charcoal; there was a continued vivid light in the galvanic circuit.

August 23, 1809.

AN EXPERIMENT TO ASCERTAIN WHETHER ANY HEAT SENSIBLE TO THE THERMOMETER IS PRODUCED BY THE ELECTRIC FLAME IN VACUO.

The jar which contained the apparatus consisted of a concave-plated mirror, so situated as to collect the light radiating from the charcoal, and to concentrate them (sic) on the bulb of a mercurial thermometer, which, together with the wires holding the two pieces of charcoal, passed through a collar of leather. No heat was apparently produced by the light excited

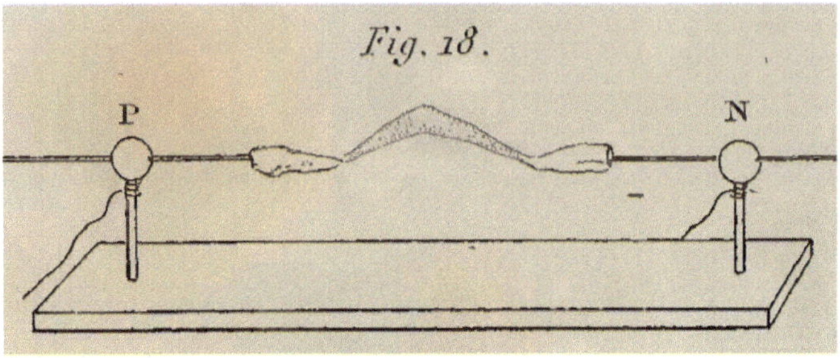

Fig. 4 Davy's drawing of an horizontal [13]

> in vacuo. The air being introduced, immediately the column of mercury rose. The light in vacuo was in part of a beautiful blue colour, and attended with bright red scintillations.

Ayrton [9] concluded that:

> The "vivid light" referred to in the first of these extracts is plainly an arc; but in the second, the words "electric flame" leave no room for doubt, not only that Davy was using an arc, but that it was no new phenomenon to him. When was the arc discovered then, and by whom?

Then, she explained that in 1812, Davy presented his *Elements of Chemical Philosophy* published in 1812 in which he not only experienced continuous arc discharge but also provide a drawing of it (see Fig. 4). Davy [13] wrote:

> When pieces of charcoal about an inch long and one sixth of an inch in diameter, were brought near each other (within the thirtieth or fortieth part of an inch, a bright spark was produced, and more than half the volume of the charcoal became ignited to whiteness, and by withdrawing the points from each other a constant discharge took place through the heated air, in a space equal at least to four inches, producing a most brilliant ascending arch of light, broad, and conical in form in the middle.[5]

According to Ayrton [9]:

> This very definite and beautiful description of the arc leaves no doubt that Sir Humphry Davy was the first to show the long horizontal arch of flame that gives the arc its name[6]; although the question whether or not he was the first person to obtain an arc of any shape and size will probably remain for ever a mystery.

It seems so that she was unaware of Petrov's works at that time. From 1812 to 1820, Davy didn't do any important work concerning the electric arc. In July 5, 1821, Davy presented his new experiments on the influence of a magnetic field on the electric arc before the members of the Royal Society of London. Davy [15] wrote:

> Mr. Pepys having had the goodness to charge the great battery of the London Institution, consisting of two thousand double plates of zinc and copper, with a mixture of 1168 parts of water, 108 parts of nitrous acid, and 25 parts of sulphuric acid, the poles were connected by

[5]Davy referred to Plate III, Fig. 18 which is presented in Fig. 4.

[6]Ayrton explained that François Arago gave to the electric flame the name of "arc". See Arago [14].

> charcoal, so as to make an arc, or column of electrical light, varying in length from one to four inches, according to the state of rarefaction of the atmosphere in which it was produced; and a powerful magnet being presented to this arc or column, having its pole at a very acute angle to it, the arc, or column, was attracted or repelled with a rotatory motion, or made to revolve, by placing the poles in different positions ...

According to Anders [8], such "column of electric light" is nothing else but a plasma.

Thus, as a conclusion of this first part, it appears that Petrov in 1803 and then Davy in 1808 have been the very first to observe continuous arc discharges, i.e., plasma. A full history of the electric arc can not be presented in this chapter, so, for more details, we refer the reader to Ayrton [9] and Anders [1, 8, 16].

A few years after this discovery, one of the main practical applications of continuous arc discharges was for street and large building lighting. Thus, the first electric street lighting employing carbon arc lamps, also called "Electric candle", was developed in 1875. On 30 May 1878, the first electric street lights in Paris were installed on the avenue de l'Opera and the Place de l'Etoile, around the Arc de Triomphe, to celebrate the opening of the Paris Universal Exposition. In 1881, to coincide with the Paris International Exposition of Electricity, street lights were installed on the major boulevards. The first streets in London lit with the electrical arc lamp were by the Holborn Viaduct and the Thames Embankment in 1878. More than 4,000 were in use by 1881. However, regardless of weak glow produced by electric arc, it had a major drawback: the noise generated by the electrical discharge which inconvenienced the population. As we will see below the solution of this problem will give rise to the birth of *Wireless Telegraphy*.

2 The Birth of Wireless Telegraphy

The historical development of radiotelegraphy has been summarized as follows by Stanley [17]:

> The development of radio-telegraphy, from the fundamental fact that under certain circumstances an electric discharge will oscillate at a high frequency, may best be explained by treating it in a historical manner (...).
>
> As already pointed out **Prof. Henry** in **1838** (and later Fedderson in 1857) discovered that under suitable circumstances the discharge of a Leyden jar, or condenser, would be oscillatory, and would oscillate at high frequency. **Kelvin** (**1853**) proved the laws under which these oscillations took place; showed that the time of oscillation depended on $\sqrt{LK}$, and that the effect of resistance was to damp the oscillations. The conception of electric and magnetic strains in the ether was due to Faraday; but Faraday, whilst a brilliant experimentalist, was not a mathematician and thus did not foresee the far-reaching results of some of his experiments. Faraday died in 1867.
>
> In **1863 James Clerk Maxwell**, by mathematical reasoning, formulated the theory that light and radiant heat were electro-magnetic phenomena, caused by strains set up in the all-pervading medium, similar to the electric lines and magnetic lines which we have already discussed. He said that electro-magnetic disturbances traveled in the ether at a definite velocity, that is to say the known velocity of light and radiant heat 186,000 miles, or 300,000

> kilometres per second. James Clerk Maxwell was a mathematician, not an experimentalist, and his theories lacked experimental proof for 24 years.
> However in **1887 Heinrich Hertz**, a young German Professor, issued the results of his experiments, which proved conclusively the correctness of Maxwell's theory (...).
> It will be remembered that when an electrical discharge takes place in an oscillatory circuit a portion of the oscillating energy is communicated to the surrounding ether [environment] in the form of electric and magnetic strains, producing in it a wave motion [the propagation of electromagnetic waves].
> Thus the arrangement of circuit made by Hertz was the first *open type of oscillator*. It consisted of a spark gap on each side of which were copper rods, 30 cms. long, terminating in large square brass plates of 40 cms. side, or round discs of copper, brass, or zinc. Such a Hertzian oscillator is shown in Fig. 5. Each side of the spark gap is joined to the high potential terminals of an induction coil, by which the oscillator is charged; the plates provide a suitable amount of capacity effect in the circuit; inductance effect is present in every circuit, even with straight wires. Using a suitable length of spark gap the discharge of this Hertzian open circuit is oscillatory. (...)
> Hertz detected the presence of ether disturbances, or ether energy [electromagnetic waves], in the space around his apparatus by using what he called a "resonator," corresponding to a receiver in radio-telegraphy. It simply consisted of a stout copper wire circle (see R in Fig. 5), of about 35 cms. radius, with a very small spark gap in the circle. When he held this circle parallel to his oscillator [O], in such a way that the small spark gap was turned towards it, he obtained minute sparks across the resonator gap.

Thus, the experiments of Hertz proved the accuracy of Maxwell's theories as well as the existence of electromagnetic waves or radio waves and opened up a new and delightful field of scientific investigation. Nevertheless, it appeared as early as 1891 that the insufficient power caused by the high damping of spark-generated waves was a barrier to overcome. So, physicists and engineers tried to construct a continuous-wave transmitter which was finally discovered almost by chance.

2.1 Duddell Singing Arc

Indeed, in London, the physicist William du Bois Duddell (1872–1917) was commissioned in 1899 by the British authorities to eliminate the noise generated by the electrical discharge of the carbon arc lamps used in the street lights. He thought up the association of an oscillating circuit made with an inductor L and a capacitor C (F on Fig. 6) with the *electrical arc* to stop the noise (see Fig. 6). Duddell [18, 19] created a device that he named *singing arc*.

Duddell had actually created an oscillating circuit capable of producing not only sounds (hence its name) but especially continuous electromagnetic waves. This device would therefore be used as an emitter for *wireless telegraphy* until the triode vacuum tube replaced it. The *singing arc* or *Duddell's arc* was indeed a "spark gap" device meaning that it produced sparks which generated the propagation of electromagnetic waves shown by Hertz's experiments as pointed out by Poincaré [20]:

> If an electric arc is powered by direct current and if we put a self-inductor and a capacitor in a parallel circuit, the result is comparable to Hertz's oscillator...These oscillations are

Fig. 5 Hertz spark gap oscillator

> *sustained* exactly like those of the pendulum of a clock. We have genuinely an electrical escapement.

In a series of "forgotten lectures" given by Poincaré [21] in 1908 at the École Supérieure des Postes et Télécommunications (today Telecom ParisTech) and recently "re-discovered" by Ginoux [5, 22–30], he stated that the existence of *sustained oscillations* in the singing arc represented a stable regime of continuous waves necessary for radio communication.

While improvements were brought to Hertz's damped oscillator with Duddell's singing arc, i.e. to transmitter of radio waves now to be called Hertzian waves (as above recalled), a new kind of detector was developed. Indeed, the small spark gap

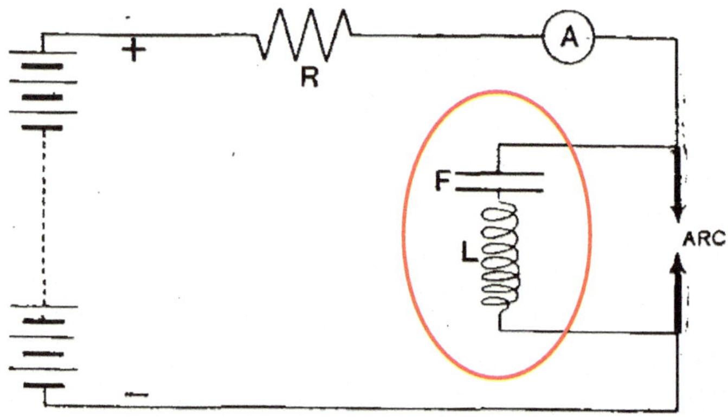

Fig. 6 Diagram of the singing arc's circuit, from Duddell [18, 19]

of Hertz in his circle of wire, was a very crude detector, making manifest sparks only at very short ranges. So, it became necessary to build a more sensitive detector which would be also able to keep a traces of the sparks, i.e. of the Hertzian waves emission. Such "memory effect device" was built as earlier as 1890 by Édouard Branly.

2.2 *Branly Coherer and Branly Effect*

Édouard Branly (1844–1940) was a French physician and physicist who greatly contributed to the development of *Wireless Telegraphy* by making one of the very first detectors of Hertzian waves. Thus, by coupling the Hertz oscillator or Duddell singing arc, i.e., a transmitter with a detector, i.e., a receiver, it became possible to build portable transmitters and receiver apparatus turning what was essentially a laboratory experiment into a useful communication system. According to Dilhac [31]:

> At the end of the 1880s, Branly went back to his research in pure physics, concentrating on the influence of irradiation on the electrical conductivity of various substances. In June 1890 he used a Wimshurt machine to create sparks and to study the electrostatic discharge of various substances submitted to the light from the spark (i.e., UV-rich radiation). Branly had devised a first circuit to create sparks using the Wimshurt machine, and then a second very simple circuit: a Daniell battery, a galvanometer, and a metallic disk all wired together in series. The disk was initially electrically charged. The two circuits were initially close to each other, so the light from the spark illuminated the disk. Following the spark, in some cases a dramatic increase in disk conductivity could be detected by the galvanometer. In November 1890 he replaced the disk by a tube filled with oxidized Zn particles. Just after the spark, it was again found that the conductivity of the tube was increased by several orders of magnitude. (...) He then put the circuit made of the battery, galvanometer, and tube in another room, 20 meters away, separated by thick walls and a courtyard from the Wimshurt spark-emitting circuit: the effect persisted while neither the light from the spark nor its sound could be seen or heard by Branly sitting by the tube while his aide Rodolphe

Gendron was operating the electrostatic machine. A small shock was found to restore the initial conductivity value,[7] while a new spark allowed the phenomenon to be repeated.

In 1891, Branly continued his research which were published in the French journal *La Lumière Électrique* [33] and translated in the English journal *The Electrician* [34]. His contribution entitled "Variations Of Conductivity Under Electrical Influence" was also reproduced in the appendix of the book of Sir Lodge [35]. Let's notice that in this book, Lodge also described, in a chapter devoted to "The History of the Coherer Principle," work prior to 1892 by a number of experimenters on the cohesion principle, but himself only became aware of Branly's work in 1892. However, in his work, Branly explained with many details his experiments and device:

The object of this article is to describe the first results obtained in an investigation of the variation of resistance of a large number of conductors under various electrical influences. The substances which up to the present have presented the greatest variations in conductivity are the powders or filings of metals. The enormous resistance offered by metal in a state of powder is well known; indeed, if we take a somewhat long column of very fine metallic powder the passage of the current is completely stopped. The increase in the electrical conductivity by pressure of powdered conducting substances is well known, and has had various practical applications. The variations of conductivity, however, which occur on subjecting conducting bodies to various electrical influences have not been previously investigated.

In the same article, Branly [33–35], described what is nowadays known as the *Branly effect*:

The Effect of Electric Sparks. - Let us take a circuit comprising a single cell, a galvanometer, and some powdered metal enclosed in an ebonite tube of 1 square centimetre cross section and a few centimetres long. Close the extremities of the tube with two cylindrical copper tubes pressing against the powdered metal and connected to the rest of the circuit. If the powder is sufficiently fine, even a very sensitive galvanometer does not show any evidence of a current passing. The resistance is of the order of millions of ohms, although the same metal melted or under pressure would only offer (the dimensions being the same) a resistance equal to a fraction of an ohm. There being, therefore, no current in the circuit, a Leyden jar (see Fig. 7) is discharged at some little distance off, and the abrupt and permanent deflection of the galvanometer needle shows that an immediate and a permanent reduction of the resistance has been caused. The resistance of the metal is no longer to be measured in millions of ohms, but in hundreds. Its conductivity increases with the number and intensity of the sparks. (…)
Restoration of Original Resistance. - The conductivity causes by the various electrical influences lasts sometimes for a long period (24 h or more), but it is always possible to make it rapidly disappear, particularly by a shock.

Let's notice that the Branly effect remained unexplained and misunderstood till the very recent studies of Eric Falcon and Bernard Castaing [36].

In this "ebonite tube" (see Fig. 8), the resistance horizontally between plates C and D, or vertically between plates A and B, could be measured independently. So, Branly found that the conductivity through his tube changed simultaneously in both directions. Dilhac recalled that in 1892:

[7]Branly [32] wrote in his note: "the variation of resistance is almost completely suppressed by various methods, notably by striking a few small yanks on the tablet which supports the tube."

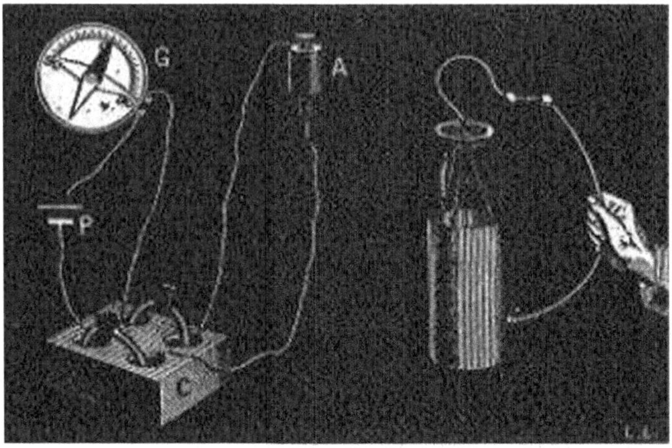

Fig. 7 Experimental setup used by Branly [33]

Fig. 8 Branly's filings tube (radioconductor or coherer), from Branly [33]

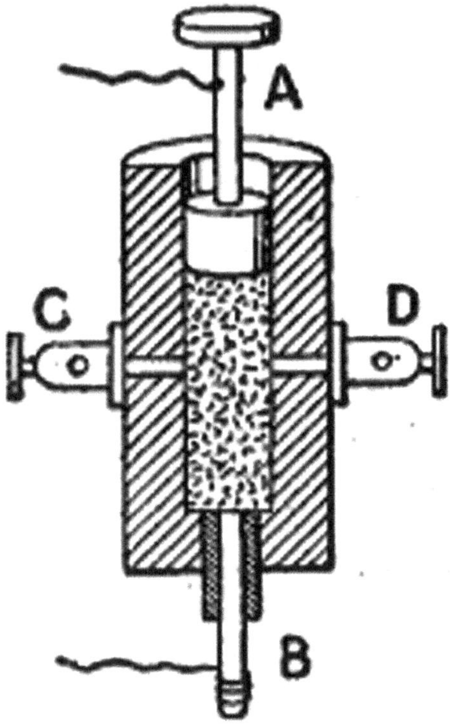

> Lodge improved Branly's tube by adding a relay automatically triggering a shock after a decrease in electrical resistance, making the device usable for wireless transmission. Lodge coined the term Branly's coherer (from Latin *cohaere*, to stick) referring to his first understanding that the modifications in resistance were related to small movements of the particles, considered as dipoles, due to electrostatic effects.

However, Branly didn't accepted the term *coherer* for his device to which he preferred the neologism *radioconductor*. In 1897, Branly [37–39] explained the reasons of his choice:

> My filings received from Lodge the name of *coherers*, this name has been generally accepted. This expression is based on an incomplete examination of the phenomenon and an incorrect interpretation; I have proposed the name of *radioconductors*, which reminds us of the essential property of discontinuous conductors to be excited by electrical radiation.

From 1894 and while using a Hertz's oscillator as transmitter and a Branly coherer as receiver, the famous Italian inventor Guglielmo Marconi (1874–1937) began to conduct experiments in radio waves. The next summer, he was able to transmit signals up to one half mile. He made his first demonstration of his transmitter-receiver system for the British government in July 1896 and in March 1897, he transmitted Morse code signals over a distance of about 3.7 miles. In France, on November 5, 1898, the scientific instrument manufacturer, Eugène Ducretet (1844–1915) performed a demonstration of wireless communication in the presence of representatives of the Acadèmie des Sciences between the third floor of the Eiffel Tower and the Panthéon 2.5 miles away while using a spark-gap transmitter of his invention and a Branly coherer as receiver. Between March 27 and 29, 1899 Marconi realized the first wireless transmission over the Channel between Dover in England and Wimereux in France with his system (see Fig. 9).

According to Stanley [17] (see Fig. 9):

> Marconi included in the local circuit a little electro-magnet [T], so that when the local current flowed the magnet attracted an armature of soft iron to which was attached a little hammer; this hammer striking against the coherer [C] caused it to decohere so that it was ready to be again affected by another impulse, or train of ether waves [electromagnetic or radio waves]. The time of action could be made long or short according to the length of time the oscillator switch was closed; thus a scheme of signals, such as the Morse Code, could be used.

On March 29, 1899 Marconi sent the following telegram to Édouard Branly:

> Mr. Marconi sends to Mr. Branly his regards over the Channel through the wireless telegraph, this nice achievement being partly the result of Mr. Branly's remarkable work.

In the beginning of the twentieth century, while using the Duddell's singing arc (see Sect. 2.1), the Danish physicist, Valdemar Poulsen (1869–1942), built an arc generator that produced *continuous waves* with very high frequencies. Then, many efforts were made to transmit human speech with such system. Nevertheless, it appeared that it was impossible with spark technology. Moreover, although the coherer was used successfully for several years, it soon showed its poor efficiency and sensitivity. It was progressively replaced by crystal of Galena also called "cat's-whisker detector" and then, by the famous Lee de Forest (1873–1961) "audion" valve or triode. This

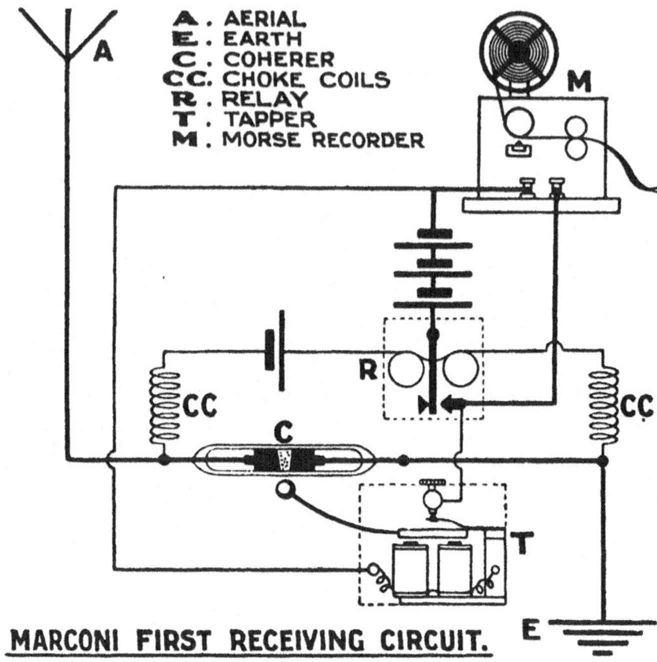

Fig. 9 Marconi's first receiving circuit, from Stanley [17]

electronic detecting or amplifying vacuum tube led to a significant development of *Wireless Telegraphy*. Then, Branly coherer and the Branly effect seemed to have been completely forgotten during nearly half a century (see next section). Thus, during the night of April 15, 1912, the *Titanic* had sent its last message with the Marconi Wireless Installation.[8] Nevertheless, a magnetic detector, commonly known as "Maggie," working in conjunction with a Marconi multiple tuner, replaced the less-efficient coherers of previous years. However, eight months after *Titanic* disaster an article entitled "Hunting for Gold with Wireless Waves" was published in July 1912. In this really incredible publication, the authors claimed that:

> As is well known, the wireless waves are not electric, but a form of oscillations or waves set up in the ether by the discharge of electric sparks. These oscillations readily penetrate paper, porcelain, sands and clays. But should a metal plate be encountered they are reflected. It is on these well demonstrated principles that the use of wireless for exploring the earth's interior is based. In practice two instruments are employed, the oscillator, or sender, and the coherer or receiver, substantially of the same nature as those used in wireless telegraphy. Each of these instruments is placed at the focal point of a parabolic metal reflector. In exploring the

[8]Marconi operators used first the distress signal - CQD - which boomed out over the Atlantic. Then, they desperately tried the new distress signal - SOS - that had been introduced a few years before. In 1906 the International Radio Telegraphic Convention in Berlin created the signal "SOS" for summoning assistance. The letters were chosen for their simplicity in Morse Code - three dots, three dashes and three dots.

earth, the oscillator and coherer are so adjusted that the waves emanating from the former are reflected so that, if no obstacle were interposed, they would strike the mirror of the latter in parallel lines and therefore converge on the coherer at its focus. But in the path of the rays from the transmitter to the receiver is placed a metal plate which prevents the passage of waves to the receiver and reflects them obliquely into the earth. The waves pass readily through comparatively dry strata of sand and clays, much as light penetrates glass, but when an ore body or watercourse is encountered the rays are deflected upward through the earth as a mirror casts back sunlight . The waves in their upward flight are arrested by the reflector of the receiver and the angle to which the receiver's reflector must be turned in order to catch and focus them on the coherer, and the distance between the center of the interposed reflector plate and the focal point of the receiver's reflector furnishes a means to calculate the .depth and position of the ore body or watercourse. An electric gong inserted in the power circuit of the receiver gives the alarm when rays pass from the oscillator to coherer.[9]

3 Coherer-Based Computer Memories

According to Dilhac [31] "With the absence of subsequent industrial applications, the lack of accuracy of the coherer as a scientific instrument, and the difficulty of elaborating a definitive theory despite the comprehensive experimental work performed by Branly, the Branly effect fell into oblivion.". Indeed, in the "Branly effect", recalled in the previous section, the falling of the resistance through the coherer tube lasts a certain time (Branly [33–35] spoke about "24 h or more"), longer than the spark produced by the electric arc. So, it can be interpreted as a "memory effect". That's probably the reason why a serious attempt was made during the early 1950s to study the feasibility of coherer-based computer memories and / or logic gates. Because some types of coherers were bistable, there was a possibility of using them as memory elements. Far from being far fetched, this idea was taken very seriously by the US government as exemplified by the four reports [40–43] discovered by one of us [44] and which will be briefly analyzed below. Thus, at the end of 1950, scientists of the Mellon Institute Of Industrial Research (founded in 1913 by Andrew W. Mellon and Richard B. Mellon and located in Pittsburgh) were hired to proceed on a research program for the study of computer components. The aim of this program was to discover and develop devices which would perform one or more of the functions of digital computers, namely storage, selection, amplification, comparison, and arithmetic operation on information pulses. Components small in size, light in weight, low in cost and capable of being manufactured on a large scale were sought. So, in the first "Quarterly Report" entitled: "Computer Components Fellowship" [40] one can read in the abstract:

> Electrolytic diode rectifiers, cold-cathode discharge tubes and coherers have been the subject or active study during the period covered by this report. Each of the devices shows potential promise as a computing machine component.

Then, in paragraph III of this reports, entitled "Coherers", the authors recall in their introduction the Branly effect:

[9] *Popular Mechanics Magazine* Volume 18, No. 6, 858–859, December 1912.

> A number of metallic powders behave as insulators under the influence[10] of a low voltage and as conductors under the action of a high voltage. A coherer is a device for making use of this property, and it consists essentially of a pair of electrodes immersed in the voltage-sensitive powder. The transition from the non-conducting to the conducting state takes place abruptly when the potential difference between the electrodes reaches a certain critical value, and the coherer remains in the conducting state even after the "firing" voltage is removed. A return to the nonconducting state is brought about by mechanical vibration or some other suitable means of erasure. The coherer may also be transferred from the conducting state under the influence of electromagnetic radiation. For example, a spark occurring at a distance of many feet from the sensitive powder is sufficient to "fire" or cohere it, and, in this laboratory, a particular coherer was made to fire simply by turning on a power supply unit in its near vicinity. The use of the coherer by Marconi to detect wireless messages is, of course, well known.
> A reliable device having the properties described above could clearly serve the functions of switching and storage in a digital computer, and for this reason it was decided to study the coherer in detail.

In their discussion, the authors concluded as follows:

> The preliminary experiments described above indicate that the coherer is a promising device for performing switching and storage functions. Its action as a switch is similar to that of a single pole-single throw relay in that the switching circuit may be isolated completely from the information carrying circuit. Its "on-off" action may be made extremely fast if the erasure is effected by "mechanical tapping" in the form of a supersonic pulse, and erasure by this means is justified if the application permits the simultaneous erasure of a large number of coherers as, for example, in a pyramid-selector fabricated from coherers or in a storage block containing a large number of digits. The lack of a rapid method for the erasure of individual coherers is therefore not a critical limitation. More serious is the lack of experience and information on the problem of producing coherers with reproducible characteristics. The parameters affecting the behavior of these devices will therefore be the subject of future study.

This last problem had been already observed by Marconi as recalled by Hong [45]:

> Marconi had difficulties with his receiver. He started with a Branly-tube coherer and a galvanometer. (...) However, he found the Branly coherer unstable: it "would act at thirty feet from the transmitter, [but] at other times it would not act even when brought as close as three or four feet."

Unfortunately, it has not been possible for instance to obtain the second "Quarterly Reports". However, in the abstract of the third one can read:

> The work on electrolytic diodes, cold-cathode gas diodes, and coherers presented in Quarterly Reports Nos. 1 and 2 of the present series has been continued. The new information on these components is set forth in the first three sections of this report.

Indeed, in paragraph III of this third reports, also entitled "Coherers", the authors present the results of their new experiments:

---

[10] By applying a radio frequency pulse directly to the coil surrounding the voltage-sensitive (100 mesh) iron powder.

> As explained in Quarterly Progress Reports Nos. 1 and 2 of the present series, certain metallic powders behave as insulators under the influence of a low voltage and as conductors under the action of a higher voltage. A coherer is a device for making use of this property, and it consists essentially of a pair of electrodes immersed in the voltage-sensitive metallic powder. On raising the d.c. potential difference between the electrodes to a certain critical value, the resistance of the coherer falls from an essentially infinite value to some value in the range from zero to one hundred ohms. For "coherer" powders selected at random the critical ("firing") voltage as well as the "fired" resistance varies by a factor of two or three on repeated operation. To effect a return to the state of infinite resistance it is generally necessary to mechanically tap the housing which contains the voltage-sensitive material. We have found, however, that certain types of coherers may be "decohered" by discharging a condenser across the coherer electrodes. The coherer may therefore be returned from a state of finite resistance to infinite resistance, and vice versa, by electrical means alone. This situation suggested the possibility of constructing a "coherer oscillator,"[11] and a low-frequency oscillator was in fact constructed. It is described in the succeeding part of this report.
> In the above discussion reference was made to "firing" coherers with the aid of a d.c. potential difference. A radio frequency voltage applied across the coherer electrodes is equally effective. In fact if one side of a capacitor is connected to a radio frequency voltage source, and the other electrode of the capacitor is dipped into the voltage-sensitive powder, the resistance between the coherer electrodes immediately falls to a low value. Also, if several electrodes are immersed in the voltage-sensitive powder; the resistance between any arbitrarily chosen pair will be found to be quite low (Quarterly Progress Report No. 2, p III-7). Coherers will, in fact, "fire" in radio frequency fields, doubtless because of the voltage induced between the coherer electrodes. The problem of "firing" coherers is therefore a simple one; the problem of decohering them is much more difficult. The capacitor method of decohering described above is believed to be a low-frequency electromechanical phenomenon, since the decohering action was observed to be frequently accompanied by disruptive sparks within the voltage-sensitive mass. With respect to individual coherers there consequently appear to be two main problems:
>
> (a) the problem of decohering and
> (b) the problem of obtaining coherers which are more uniform in regard to firing potentials and cohered resistance values.
>
> Research pointed toward the solution of these problems is justified only if "ganged" coherers may be used as successful switching devices, and we describe for the first time in part 3 of the present section a successfully operating binary to octal digital converter using coherers as the sensing-elements. Part 4 of the present section describes certain unsuccessful attempts to effect decoherence by means of supersonic pulses and magnetic phenomena.

So, as soon as, 1951, a binary-to-octal converter employing coherers as the switching elements was considered. This attempt is described in the same third "Quarterly Report" [41] as follows:

> In Quarterly Progress Report No. 2 some attempts to switch with three electrode coherers were described. Two of the electrodes consisted simply of wires immersed in a voltage-sensitive powder. The third electrode (immersed in the powder) consisted of one terminal of a capacitor, the other end of which could be switched to a radio frequency voltage source. In this way the coherer was "fired"; the resistance between arbitrarily chosen pairs of the

---

[11] More than forty years before, Eccles [46] provided in a contribution published in 1909 the set of ordinary differential equations modeling the coherer's oscillations.

three electrodes falling from a very high value ($>10^9$ ohms) to only a few ohms. For single isolated coherers the "firing" technique appeared to be 100% efficient. However, when two or more coherers were connected in series, an r.f. pulse applied to one of the coherers fired not only the immediately affected coherer, but one or more of the neighboring ones. This behavior was eliminated almost entirely by connecting the junction points between coherers to ground potential through small capacitors. An effort was therefore made to adapt this switching technique to the construction of a binary-to-octal digital converter. This more complicated set-up unfortunately did not work in a reproducible way.

In the fourth "Quarterly Report" [42], the authors thus conclude that:

Because of their bi-stable nature, coherers are natural devices for the storage of binary information. They may also be used as "switching elements" as demonstrated in Quarterly Progress Report No. 3 where a binary-to-octal digital converter employing 14 coherers as the switching elements was described. The application of coherers as either storage or switching elements is, however, considerably vitiated by the need for mechanical agitation to effect decoherence, that is, to effect return from the state of low resistance to (practically) infinite resistance. In the Progress Report referred to above, an account of several unsuccessful attempts to bring about decoherence by means of supersonic waves and by magnetic fields was described. These experiments are continuing as time permits but no noteworthy new results have been obtained.

In their fifth "Quarterly Report" [43], they give some of the reasons for which they have probably chosen to study the "coherer":

The coherer consists of an aggregate of oxide-coated powder in which are immersed two metal electrodes. The device is of 1nterest because it is a bi-stable element which will remain in either of its two stable states without the use of any "holding power". In addition, it is extremely inexpensive. The information contained in the present report throws considerable light on the mechanism of its operation. (...)
The section on coherers in this report is devoted largely to the behavior characteristics of this device. In particular, the mechanisms of coherence and decoherence are examined at some length, and the implications of the findings for the potential application of the device as a computer component are discussed. The basic facts are these: the coherer constitutes an attractive device for computer application because it may be maintained in either of two stable states without the use of an external power source. It suffers from the disadvantage that it is unsymmetrical in operation in that the time (energy) required to transfer it from a state of high resistance to a state of low resistance (coherence) is much smaller than the time (energy) required to effect the reverse transition (decoherence). The suitability of the device as a computer component increases as the time required for decoherence decreases. The major effort at the present time is therefore pointed toward decreasing the decoherence time. Decoherence by means of magnetic forces shows some promise.

Then, they presented the "Mechanisms of Coherence" or Branly effect as follows:

The voltage-sensitive powder employed in coherers consists generally of finely divided metal particles each of which is coated with a very thin oxide layer. When these particles are placed in contact with each other in an electrical field, a redistribution of charge is effected such that the electrical field within the particles proper is reduced to zero. (...)
As the potential difference between the electrodes is slowly increased a point will be reached at which one or more of the oxide layers will break down and bridge the gap between the adjacent particles with a filament of metal. This action will have the effect of raising the potential difference across the remaining oxide interfaces and thus lead to further breakdown. The process is therefore of a cascade type and occurs in an extremely short time.

The mechanism described above corresponded to the most well-known theory at that time concerning the action of the coherer. This was that of Holm [47] quoted by the authors of this fifth "'Quarterly Report" a few pages below. Finally, they discuss the possibility to use the "coherer as a computer component":

> As mentioned in the introduction to the present section, the coherer is attractive as a computer component because it remains in either of its stable states without the application of power. Its future as a computer component depends primarily on the speed with which it may be transferred from one state to another, and more particularly on the rate at which it may be decohered. The evidence described above indicates that if a powder-mass coherer is to be used in a reproducible way, considerable energy must be packed into the decohering process, and this energy must impart an impulsive movement to the cohered chain. Attempts to supply the energy magnetically have been unsuccessful until recently because, in previous experiments, the mass motion of the metal powder was inhibited and constrained. If the cohered mass is, however, permitted some freedom of motion, and if the holding-current is released, decoherence will result.

A measure of the importance of these "Quarterly Reports" can be appreciated from the fact that no less a leading computer designer at the time than John Presper Eckert (1919–1995), the co-inventor of the ENIAC (Electronic Numerical Integrator and Computer), one of the earliest electronic general-purpose computers made, referenced these reports in one of his papers [48]. Moreover, these researches concerning the coherer were far from being the only one. From the end of September 1953, at least four confidential IBM report, recently declassified were written by a systems engineer named Fred Bernard Wood (1917–2006). In the first report entitled "The coherer as a storage element", Wood [49] wrote in his conclusions:

> The increase of the mean deviation of the firing voltage indicates potential troubles with multiple coherer units. In view of these potential difficulties, it is recommended that the coherer action be more thoroughly explored and checked by suitable experiments.

On April 28, 1954, in his second confidential IBM report, Wood [50] provided an extensive chronological biography on coherers. We only reproduce below the interesting period of 1910–1953 with Wood's annotations:

176. B. Szilard, "Action of Metallic Contacts on a Filings Coherer."
Compt. Rend. 150, 1670–1672 (1910); Sci. Abst. A13.1133 (1910).
177. E. C. Green, "The Development of the Coherer and Some Theories of Coherer Action." General Electric Review, 20, 369–374 (1917).
178. E. C. Green, "Development of the coherer and some theories of coherer action." Sci. Am. S. 84, 268–9 (Oct. 27, 1917).
179. W. G. Palmer, "The Use of the Coherer to Investigate Adsportion [*sic*, should read "Adsorption"] Films."
Proc. Roy. Soc. A. 106, pp. 55–68 (1924).
Study of the effect of the gas surrounding the "coherer" detector.
concludes that adsorbed gas film is displaced by high enough field.
180. R. H. Wright and M. J. Marshall, "The Effect of Adsorbed Gas on the Contact Resistance of Carbon."
Trans. Am. Electrochem. Soc. 454 [*sic*, should read 54], 149–162, Sept. 1928.
Oxygen is adsorbed on carbon surfaces which changes the contact resistance.
This may help explain carbon coherer.

181. John R. Bowman, "Electrochemical Computing Elements." Annals Comp. Lab. Harvard, 119–124 (1951); Proceeding of a Second Symposium on Large-Scale Digital Calculation Machinery, Sept. 1949.
182. Harvard Computation Laboratory, "Electrochemical Computer Elements." Progress Report No. 8, pp. XII-1 to XII-10.
183. Ragnar Holm, "The Electric Tunnel Effect across Thin Insulator Films in Contacts." Jour. Appl. Phys. 22, 569–574 (1951); Errata, 1217.
184. Mellon Institute of Industrial Research, Quarterly Reports of the Computer Component Fellowship No. 347. Contract CLN AF 19 11221–1236 J. Coherer research is reported in reports Nos. 1, 2, 3, 4, and 5 (1951–1952) and No. 12 (1953).

Wood's bibliography on coherers is of great importance since it shows that, contrary to what one thought, the coherer had not been "completely forgotten" after the advent of the vacuum tube diode or triode in 1910s. Thus, the only gap in its history (which should be subject to extensive research) seems to be between 1928 and 1946. Nevertheless, even during this period, Holm [47] had made many investigations in this field as exemplified by the bibliography of his book. Moreover, the last reference of the above bibliography proves that Wood was aware of the works performed at the Mellon Institute of Industrial Research. Wood also listed the "IBM Coded Reports on Coherers and Related Subjects"

- 100.000.239 American Phys. Soc., Rochester, June 18, 1953. Abstract of P. Kisliuk, "Electrical Breakdown of Extremely Short Gaps."
- 102.003.250 C. A. Speicher, "Dielectric Rupture and Repair Storage."
- 203.001.026 H. E. Singhaus, "Preliminary Progress Report on the Coherer Investigater Program." May 1, 1953.
- 203.002.047 F. B. Wood, "The Coherer acts a Storage Element." Sent. 22, 1953.
- 203.003.059 J. L Masterson and P. L. Pecchenino, "Matrix Storage Using Coherer Cells."
- 216.082.048 H. E. Singhaus, "Some Experimental Data on Coherers. Showing Cohering Voltages as a Function on Electrode Spacing and Configuration." Oct. 15, 1953.

In his third and fourth confidential IBM reports published on May 14, 1956, Wood [51, 52] concluded that:

> Separately connected coherers have regular probability of firing curves from which buffer storage coherers can be designed for a specified accuracy, provided the frequency of erasing information is only a few times a second. The probability of error is decreased by putting three cells in parallel. Coherer matrices without back circuit eliminators are theoretically possible, but have not been satisfactorily realized. The failure to realize satisfactory matrix operation lies primarily in the lack of control of the oxide thickness and rate of growth. "Resistox[12]" treatment of coherer powder has decreased the variation of coherer characteristics with time.

[12]A treatment developed by the Glidden Co., known as "Resistox" for the preservation of copper powder has been found useful in maintaining a consistent oxide layer. Copper powder obtained from Metals Disintegrating Co. is reduced in hydrogen to remove the original layers of oxide.

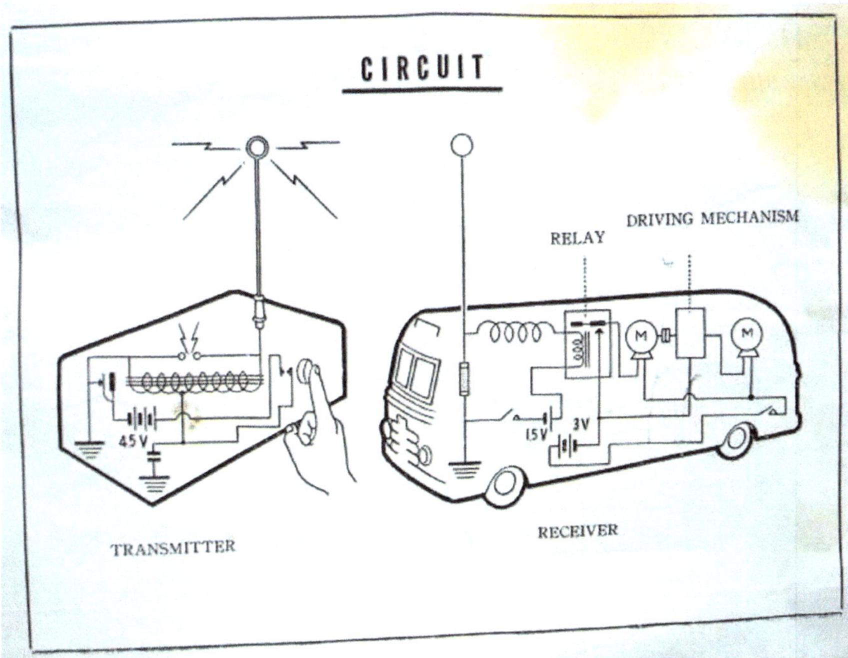

Fig. 10 Schematic of transmitter and receiver (coherer) housed in the toy bus

On September 1st, 1957, a curious application of Branly coherer was presented in *Electronics* magazine, a trade journal that covered the radio industry and its later spin-offs. It was published by McGraw Hill until 1988. At page 200, the author, probably David A. Finlay, presented a "Radio-Controlled Toys Use Spark Gap" (see Figs. 10 and 11). He explained that:

> A LINE of radio-controlled TOYS manufactured by the Masudaya Toys Co., Tokyo, Japan, goes back to one of the oldest methods of radio transmission and reception, the spark gap and coherer.
> Radiation from the spark gap is wideband, 150 kc to 180 mc, and prone to cause interference with radio and tv reception, but the radiated signal is very weak. (…)
> The bus receiver uses a coherer, a glass envelope filled with carbon powder. The antennas indicated in Fig. 1 (Fig. 10) are short vertical whips. The drive mechanism of the bus is mechanically arranged so that successive pulses cause the bus to start, turn right, go straight, turn left, go straight and stop.

4 Branly Coherer: The Very First Memristor

In 1971, Leon Chua postulated the existence of a missing electrical element, the "memristor", which was finally discovered ten years ago by Strukhov et al. [53] on May 1st, 2008. Contrary to what one might think, it is not by experimenting, but by

Fig. 11 Radicon Bus

logical deduction that Professor L. O. Chua was able to postulate the existence of a missing circuit element. In his now famous publication of 1971, [54] considered the three basic building blocks of an electric circuit: the capacitor, the resistor and the inductor as well as the three laws linking the four fundamental circuit variables, namely, the electric *current* i, the *voltage* v, the *charge* q and the *magnetic flux* φ. Then, [54] explained that:

> By the *axiomatic* definition of the three classical circuits elements, namely, the *resistor* (defined by a relationship between v and i), the *inductor* (defined by a relationship between φ and i), and the *capacitor* (defined by a relationship between q and v). Only one relationship remains undefined, the relationship between φ and q.

He thus concluded from the logical as well as axiomatic points of view, that it is necessary, for the sake of *completeness*, to postulate the existence of a fourth circuit element to which he gave the name *memristor* since it behaves like a nonlinear resistor with memory. Then, he established the "relationship" between the magnetic flux φ and the charge q:

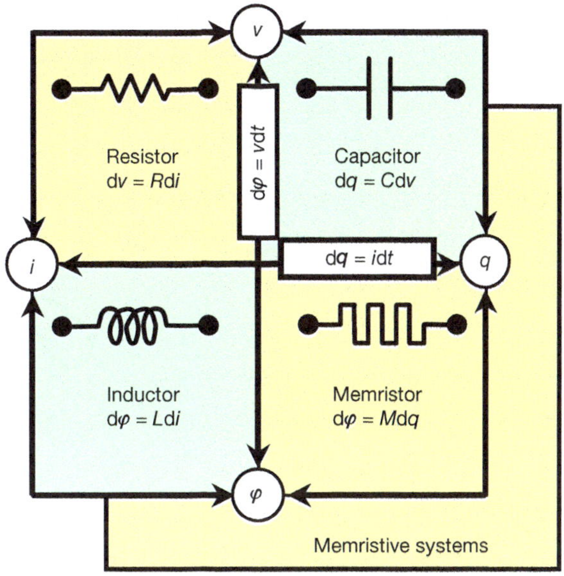

Fig. 12 The four fundamental two-terminal circuit elements, from Strukhov [53]

$$\varphi = Mq$$

We suggest to call this relationship: Chua's law. At that time, he also proposed a symbol (see Fig. 12).

In order to emphasize a strong analogy between the memristor properties and that of Branly coherer let's focus on the interesting properties 4 & 5 established by Prof. Chua in his second publication on this subject [55]. According to Chua [55] memristive systems are hysteretic:

> *Property 4–Double-Valued Lissajous Figure Property*
> A current-controlled memristive one-port under periodic operation with $i(t) = Icos\omega t$ always gives rise to a $v - i$ Lissajous figure (see Fig. 13) whose voltage v is at most a double-valued function of i

Property 5 was summarized by Chua and Kang 1976 by saying that "the $v - i$ curve is odd symmetric with respect to the origin.". In their conclusion Chua and Kang [55] gave more details about this fundamental property:

> Various generic properties of memristive systems have been derived and shown to coincide with those possessed by many physical devices and systems. Among the various properties of memristive systems, the frequency response of the Lissajous figure is especially interesting. As the excitation frequency increases toward infinity, the Lissajous figure shrinks and tends to a straight line passing through the origin-except for some pathological cases where the bibs[13] stability property is not satisfied. The physical interpretation of this phenomenon is that the system possesses certain inertia and cannot respond as rapidly as the fast variation in the excitation waveform and therefore must settle to some equilibrium state. **This implies**

[13] Bounded-input bounded-state.

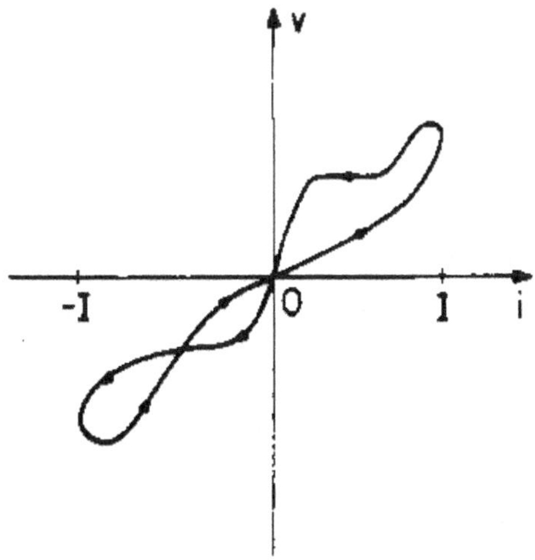

Fig. 13 Hysteretic effect, from Chua [55]

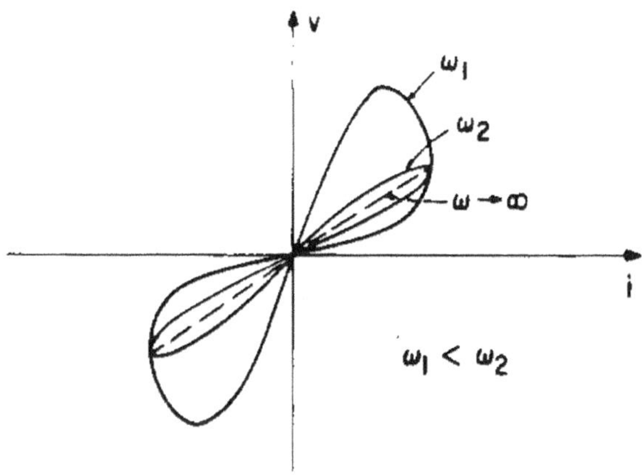

Fig. 14 Frequency response of Lissajous figures, from Chua [55]

> **that the hysteretic effect of the memristive system decreases as the frequency increases** (see Fig. 14) and hence it eventually degenerates into a purely *resistive* system.

In a recent publication, Chua [56] renamed such double-valued Lissajous figures presented in Figs. 13 and 14 as *pinched hysteresis loops*. Then, he explained that:

> All 2-terminal non-volatile memory devices based on *resistance switching are memristors*, regardless of the device material and physical operating mechanisms. They all exhibit a distinctive "fingerprint" characterized by a *pinched hysteresis loop* confined to the first and the third quadrants of the v-i plane whose contour shape in general changes with both the amplitude and frequency of any periodic "sine-wave-like" input voltage source, or current source. In particular, the pinched hysteresis loop shrinks and tends to a straight line as frequency increases.

Thus, the existence of a double-valued Lissajous figure, i.e., of a *pinched hysteresis loops* is the fingerprint of a memristor or a memristive device [57].

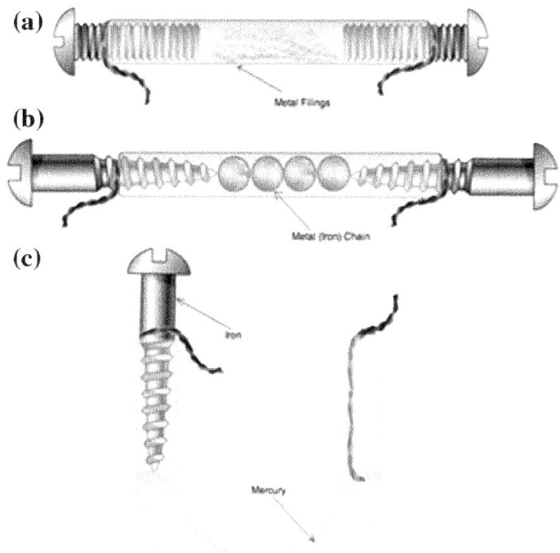

Fig. 15 Various embodiments of coherer used for experimentation, from Gandhi et al. [60]

Five years ago Gaurav Gandhi, Varun Aggarwal and Leon Chua published two articles [58, 59] in which they stated that the "Coherer is the elusive memristor". In their works, they replicated three embodiments of the coherer and autocoherer: an Iron Filing Coherer (IFC), an Iron Chain Coherer (ICC), and an Iron Mercury Coherer (IMC).[14] Their first embodiment, namely, Iron Filing Coherer (IFC), consisted of a tube containing closely-packed iron filings with electrodes in contact with the metal filings at the two ends of the tube. In the second embodiment, called Iron Chain Coherer (ICC), iron filings were replaced by a chain (linear assembly) of iron beads and the third embodiment was an embodiment of the self-recovering coherer consisting of a U-tube filled with mercury forming contact with an iron screw on one side. In the third embodiment, henceforth referred as Iron Mercury Coherer (IMC), one electrode was connected to an iron screw, whereas the other dips into mercury on the other side of the U-tube. Depending on the packing density (IFC), pressure applied (ICC) and contact area (IMC), the devices showed a continuum of states between a nonlinear high-resistance state and a more linear low-resistance state (see Fig. 15).

In the beginning of the twentieth century, the Indian physicist and physiologist, Jagadish Chandra Bose (1858–1937) made deep investigations and many experiments on the coherer. He first presented his results to the 71st Meeting of the *British Association for the Advancement of Science*, Section A at Glasgow on September 12,

[14]They repeated the experiment with several metals, including aluminum and magnesium flakes and nickel and zinc-coated ball bearings. They reported qualitatively similar results in all these experiments.

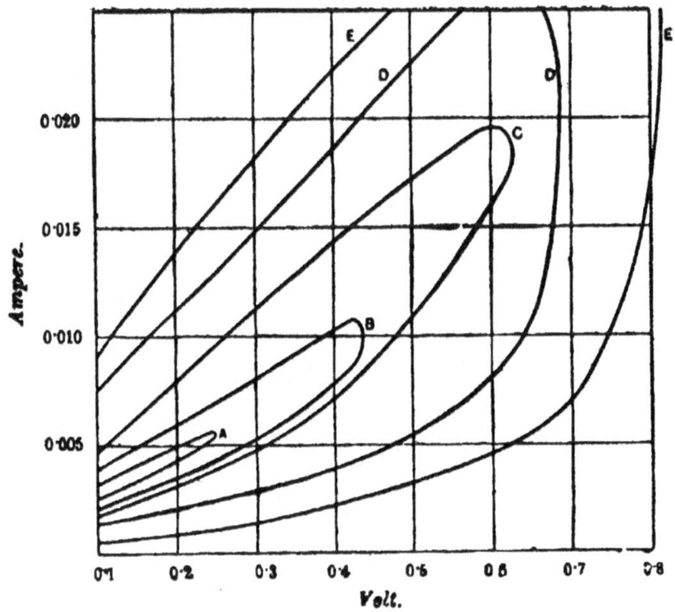

Fig. 16 Cyclic curves showing conductivity *Hysteresis*, from Bose [61]

1901. Although Bose [61] observed (probably for the first time) that coherers characteristics demonstrate a *pinched hysteresis i-v curve* in the first quadrant (see Fig. 16) and exhibited multiple stable resistance-states, he could not establish a systematic way to electrically reverse the diminution of resistance.

Then, Gandhi et al. [58–60] activated their three devices "by different current-mode input signals in their nonlinear mode, and their transient behavior was recorded." They found that the three devices show similar qualitative behavior but exhibits state-dependent resistance. They plotted the current-voltage V-I characteristics of their devices for a current-mode sine wave signal of increasing amplitude (see Fig. 17). They observed that the three devices exhibit the famous *pinched hysteresis loops* and various possible current-voltage values for the same current (as predicted by Chua [54, 55], see Figs. 13 and 14). Thus, Gandhi et al. [58–60] stated that their three embodiments of the coherer and autocoherer are *memristors*. Let's notice that such current-voltage characteristic curves had already been obtained by Falcon and Castaing [36] in 2005. Nevertheless, at that time, the *memristor* had not been already built by Strukhov et al. [53] who published their results on April 30, 2008.

So, Gandhi et al. [58–60] concluded as follows:

> On activating by any bipolar current input, the device gets programmed into one state in the positive cycle, and a different state in the negative cycle (see Fig. 17). It keeps oscillating between these two stable states, forming the famous eight-shaped pinched hysteresis loop in its V-I characteristics.

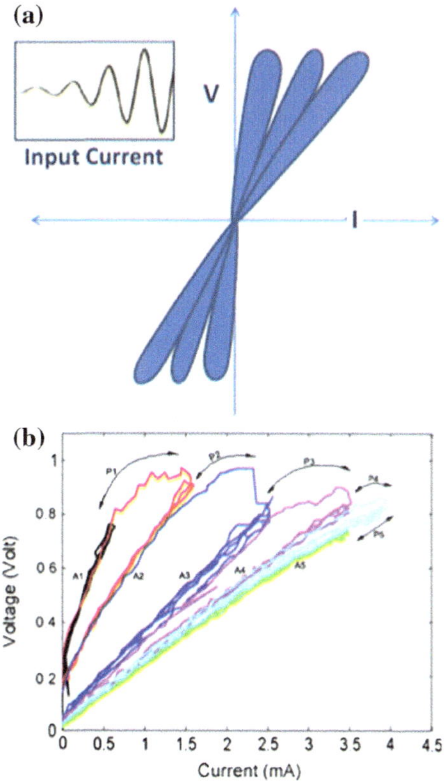

Fig. 17 Current-voltage characteristics of the three devices for a sine wave input signal, from Gandhi et al. [58–60]

> We find that the resistance of the device is a function of the magnitude of I_{max} for either directions of current, but with a quantitatively different state-map, making it behave as a resistive RAM.[15] The bistable switching,[16] the existence of the pinched hysteresis loop and ability to act like a resistive RAM qualifies the coherer and auto-coherer to be a memristor [57].

According to many researchers such as Strukhov [53] and Statholpoulos et al. [62], ReRAM (resistive Random Access Memory) is seen as a replacement technology for DRAM (Dynamic Random Access Memory) and flash memory. Resistive random-access memory (ReRAM or RRAM) is a type of non-volatile (NV) random-access (RAM) computer memory that works by changing the resistance across a dielectric solid-state material. Moreover, in a report published online in 2011, Williams claimed that "resistive RAM are both types of memristor[17]". Nowadays, memristors

[15] Resistive Random Access Memory.

[16] According to what has been recalled above, the bi-stability of some types of coherers led to the possibility of using them as memory elements.

[17] "HP and Hynix to produce the memristor goods by 2013," 10 Oct. 2011.

are intended to be used as memory elements in computer machines. So, this is not really surprising to notice that half a century ago, scientists had already considered to do the same with Branly coherer.

5 Conclusion

In this work, starting from historical and recent research, we have first recalled the origin of arc plasma science or continuous arc discharges and one of its most important applications: *Wireless Telegraphy*. Then, a summary of the development of radiotelegraphy has reminded the famous Hertz's experiments allowing on the one hand to confirm the existence of electromagnetic waves predicted by Maxwell and, on the other hand, the building of the first transmitter of radio waves now called Hertzian waves. The conception of one of the very first receivers imagined by Édouard Branly, namely the "coherer" was then presented. Although the coherer has been almost immediately superseded by diodes and triodes, its working principle, the so-called "Branly effect", gave rise to extensive investigations till very recently. In fact, subjected to electromagnetic waves, the coherer's electrical resistance varied from high to low and persisted after the radio signal was removed. Such bistable behavior led to consider the coherer as a memory device. The recent discovery of confidential reports such as the five "Quarterly Reports" as well as the four confidential IBM reports, recently declassified, has changed our point of view concerning the fact that the coherer has not been "completely forgotten" after the advent of the vacuum tube or triode in 1910s. Moreover, the brief analysis of these reports has been sufficient to show that many attempts were made during the early 1950s to study the feasibility of coherer-based computer memories. Thus, the particular features of Branly coherer, i.e., the Branly effect led naturally to associate such a device with the so-called "memristor" the existence of which has been predicted by Leon Chua in 1971. Ten years age, Resistive random-access memory (ReRAM) were considered as memristors which are intended to be used as memory elements in computer machines. Finally, five years ago, a connection between both coherer and memristor was established. So, this confirms that the many searchers and engineers who attempted to build, in the 1950s, coherer-based computer memories were not on the wrong track. To conclude, we could also say that plasma science has had a role of great importance in the development of *Wireless Telegraphy* and in that of all the devices that have been built since.

References

1. A. Anders, Tracking down the origin of arc plasma science I. Early pulsed and oscillating discharges. IEEE Trans. Plasma Sci. **31**(5), 1052–1059 (2003)
2. Sur la lumière du baromètre [On the light of the barometer], Histoire de l'Académie Royale des sciences de Paris, **2**, 202–203 (1694)

3. C. Dorsman, C.A. Crommelin, The invention of the Leyden jar. Communication n° 97 from the National Museum of the History of Science Leyden, 276–280 (1957)
4. A. Volta, *Collezione dell' Opere del Cavaliere Conte Alessandro Volta*, 3 Vol (Firenze, 1816)
5. J.-M. Ginoux, B. Rossetto, The singing arc: the oldest memristor? in *Chaos, CNN, Memristors and Beyond: A Festschrift for Leon Chua*, ed. by A. Adamatsky, G. Chen (World Scientific Publishing, Singapore, 2013), pp. 494–507
6. A. Volta, On the electricity excited by the mere contact of conducting substances of different kinds. In a letter from Mr. Alexander Volta, F. R. S. Professor of Natural Philosophy in the University of Pavia, to the Rt. Hon. Sir Joseph Banks, Bart. K. B. P. R. S. Phil. Trans. R. Soc. Lond. **90**, 430–431 (1800)
7. A.M. Ampère, Mémoire sur les effets des courants électriques. Annales de Chimie et de Physique **15**, 59–74 and 170–218 (1820)
8. A. Anders, Tracking down the origin of arc plasma science II. Early continuous discharges. IEEE Trans. Plasma Sci. **31**(5), 1060–1069 (2003)
9. H. Ayrton, *The Electric Arc* (The Electrician Printing and Publishing Company, London, 1902)
10. H. Davy, An account of some experiments on galvanic electricity made in the theatre of the royal institution. J. R. Inst. **I**, 166–182 (1802) reprinted in *Collected Works of Sir Humphry Davy*, Vol. II: Early Miscellaneous Papers, ed. by J. Davy (Smith, Elder and Cornhill, London, 1839)
11. V.P. Kartsev, V.V. Petrov's hypothetical experiment and electrical experiments of the 18th century, in *Nature Mathematized: Historical and Philosophical Case Studies in Classical Modern Natural Philosophy*, The Western Ontario Series in Philosophy of Science, vol. 20(1), ed. by W.R. Shea (D. Reidel Publishing Company, Dordrecht), pp. 279–289 (1983)
12. V.V. Petrov, *News of the Galvani-Voltaic Experiments Which Professor of Physics Vasily Petrov had Conducted by Means of a Particularly Huge Battery Consisting at Times of 4200 Copper and Zinc Disks and Installed at St. Petersburg Medicine and Surgery Academy (in Russian)* (St. Petersburg's Medical and Surgical Academy, St. Petersburg, Russia, 1803)
13. H. Davy, *Elements of Chemical Philosophy*, Part I, vol. I (J. Johnson, London, 1812)
14. F. Arago, Expériences relatives à l'aimantation du fer et de l'acier par l'action du courant voltaïque. Annales de Chimie et de Physique **15**, 93–102 (1820)
15. H. Davy, Farther researches on the magnetic phenomena conducting powers and temperature the properties of electrified bodies in their relations to produced by electricity; with some new experiments on the properties of electrified bodies in their relations to conducting powers and temperature. Philos. Trans. R. Soc. Lond. **111**, 425–439 (1821)
16. A. Anders, *Cathodic Arcs: From Fractal Spots to Energetic Condensation*, Springer Series on Atomic, Optical, and Plasma Physics, vol. 50 (Springer, New York, 2008)
17. R. Stanley, *Wireless Telegraphy*, vol. 1 and 2 (Longmans, Green & Co, New York, 1919)
18. W. du Bois Duddell, On rapid variations in the current through the direct-current arc. J. Inst. Electr. Eng. **30**(148), 232–283 (1900)
19. W. du Bois Duddell, On rapid variations in the current through the direct-current arc. The Electrician, **46**, 269–273 and 310–313 (1900)
20. H. Poincaré, *La théorie de Maxwell et les oscillations hertziennes : la télégraphie sans fil*, 3e éd. (Gauthier-Villars, Paris, 1907)
21. H. Poincaré, Sur la télégraphie sans fil. La Lumière Électrique, II**4**, 259–266, 291–297, 323–327, 355–359 and 387–393 (1908)
22. J.-M. Ginoux, L. Petitgirard, Poincaré's forgotten conferences on wireless telegraphy. Int. J. Bifurc. Chaos **20**(11), 3617–3626 (2010)
23. J.-M. Ginoux, *Analyse mathématique des phénomènes oscillatoires non linéaires, Thèse* (Université Pierre & Marie Curie, Paris VI, 2011)
24. J.-M. Ginoux, C. Letellier, Van der Pol and the history of relaxation oscillations: toward the emergence of a concepts. Chaos **22**, 023120 (2012)
25. J.-M. Ginoux, Self-excited oscillations: from Poincaé to Andronov, Nieuw Archief voor Wiskunde (New Archive for Mathematics) journal published by the Royal Dutch Mathematical Society (Koninklijk Wiskundig Genootschap), 5, **13**, n^3, 170–177 (September 2012)

26. J.-M. Ginoux, R. Lozi, Blondel et les oscillations auto-entretenues. Arch. Hist. Exact Sci. 1–46 (17 May 2012)
27. J.-M. Ginoux, *Histoire de la théorie des oscillations non linéaires* (Hermann, Paris, 2015)
28. J.-M. Ginoux, *History of Nonlinear Oscillations Theory*, Archimede, New Studies in the History and Philosophy of Science and Technology (Springer, New York, 2016)
29. J.-M. Ginoux, From nonlinear oscillations to chaos theory, in *The Foundations of Chaos Revisited: From Poincaré to Recent Advancements, Understanding Complex Systems*, ed. by C. Skiadas (Springer, Cham, 2016), pp. 27–47
30. J.-M. Ginoux, *History of Nonlinear Oscillation Theory in France (1880–1940)* Archimedes Series, vol. 49 (Springer International Publishing, Berlin, 2017), p. 381
31. J.-M. Dilhac, Edouard Branly, the Coherer, and the Branly effect [History of Communications]. IEEE Commun. Mag. **47**(9), 20–26 (2009)
32. E. Branly, Variations de conductibilité sous diverses influences électriques. Comptes Rendus de l'Académie des Sciences **111**, 785–787 (1890)
33. E. Branly, Variations de conductibilité sous diverses influences électriques. La Lumière Électrique **40**(20), 301–309 and 506–511 (1891)
34. E. Branly, Variations of conductivity under electrical influence. The Electrician **XXVII**, 221–222 and 448–449 (1891)
35. O.J. Lodge, *Signalling Through Space Without Wires, The Work of Hertz & His Successors* (D. Van Nostrand Company, London, 1908)
36. E. Falcon, B. Castaing, Electrical conductivity in granular media and Branly's coherer: a simple experiment. Am. J. Phys. **74**(4), 302–306 (2005)
37. E. Branly, Sur la conductibilité électrique des substances conductrices discontinues, à propos de la télégraphie sans fil. Comptes Rendus de l'Académie des Sciences **125**, 939–942 (1897)
38. E. Branly, Radioconducteurs limailles d'or et platine. Comptes Rendus de l'Académie des Sciences **127**, 1206–1207 (1898)
39. E. Branly, Accroissements de Résistance des Radioconducteurs. Comptes Rendus de l'Académie des Sciences **130**, 1068–1071 (1900)
40. J.R. Bowman, F.A. Schwertz, R.T. Steinback, B.O. Marshall Jr. Computer Components Fellowship No. 347, Quarterly Reports No. 1 (October 11, 1950, to January 11, 1951), *Air Force Cambridge Research Laboratory, Mellon Institute Of Industrial Research*, University of Pittsburgh, Pittsburgh, PA, Contract No. CLN AF 19/122/-376
41. J.R. Bowman, F.A. Schwertz, B. Moffat, R.T. Steinback, B.O. Marshall Jr. Computer Components Fellowship No. 347, Quarterly Reports No. 3 (April 11, 1951, to July 11. 1951) *Air Force Cambridge Research Laboratory, Mellon Institute Of Industrial Research*, University of Pittsburgh, Pittsburgh, PA, Contract No. CLN AF 19/122/-376
42. J.R. Bowman, F.A. Schwertz, A. Milch, B. Moffat, R.T. Steinback, L. Nickel, B.O. Marshall Jr. Computer Components Fellowship No. 347, Quarterly Reports No. 4 (July 10, 1951 to Oct. 10, 1951) *Air Force Cambridge Research Laboratory, Mellon Institute Of Industrial Research*, University of Pittsburgh, Pittsburgh, PA, Contract No. CLN AF 19/122/-376
43. J.R. Bowman, F.A. Schwertz, A. Milch, B. Moffat, R.T. Steinback, L. Nickel, B.O. Marshall Jr. Computer Components Fellowship No. 347, Quarterly Reports No. 5 (Oct. ll, 1951 to Jan. 10, 1952) *Air Force Cambridge Research Laboratory, Mellon Institute Of Industrial Research*, University of Pittsburgh, Pittsburgh, PA, Contract No. CLN AF 19/12.2/-376
44. T. Cuff, *Coherers, A Review* (M.S.E. Temple University, Philadelphia, 1993)
45. S. Hong, *Wireless, From Marconi's Black-Box to the Audion*, Transformations: Studies in the History of Science and Technology (MIT Press, Cambridge, MA, 2001), 272 p
46. W. Eccles, On coherers. Proc. Phys. Soc. Lond. **22**, 289–312 (1909)
47. R. Holm, *Electric Contacts, Stockholm* (Almquist & Wiksells Akademiska Handbocker, Hugo Gebers Forlag, 1946)
48. J.P. Eckert, A survey of digital computer memory systems. Proc. IEEE **41**, 1393–1406 (1953)
49. F.B. Wood, The coherer as a storage element, IBM Confidential Report, IBM Research and Development Laboratory, Code 203.002.047 (22 September 1953)

50. F.B. Wood, Coherer bibliography, IBM Confidential Report, IBM Research and Development Laboratory, Code 203.001.063 (28 April 1954)
51. F.B. Wood, Coherers characteristics part I, introduction, physical phenomena, probability and time delay, IBM Confidential Report, IBM Research and Development Laboratory, Code 203.142.102 (11 May 1956)
52. F.B. Wood, Coherers characteristics part II, materials, life test, reliability and applications, IBM Confidential Report, IBM Research and Development Laboratory, Code 203.143.107 (14 May 1956)
53. D.B. Strukhov, G.S. Snider, G.R. Stewart, R.S. Williams, The missing memristor found. Nature **453**, 8083 (2008)
54. L.O. Chua, Memristor - the missing circuit element. IEEE Trans. Circuit Theory **18**(5), 507–519 (1971)
55. L.O. Chua, S.M. Kang, Memristive devices and systems. Proc. IEEE **64**, 209–223 (1976)
56. L.O. Chua, Resistance switching memories are memristors. Appl. Phys. A **102**, 765–783 (2011)
57. S.P. Adhikari, M.P. Sah, L.O. Chua, Three fingerprints of memristor. IEEE Trans. Circuit Syst. **60**, 3008–3021 (2013)
58. G. Gandhi, V. Aggarwal, L.O. Chua, The first radios were made using memristors!. IEEE Circuits Syst. Mag. **13**(2), 8–16 (2013)
59. G. Gandhi, V. Aggarwal, L.O. Chua, Coherer is the elusive memristor, *IEEE International Symposium on Circuits and Systems (ISCAS), Melbourne, Victoria, Australia*, 2245–2248 (June 1–5, 2014)
60. G. Gandhi, V. Aggarwal, Canonic memristor: bipolar electrical switching in metal-metal contacts, in *Advances in Memristors, Memristive Devices and Systems Studies in Computational Intelligence*, vol. 701, ed. by S. Vaidyanathan, C. Volos (Springer, Cham, 2017), pp. 263–273
61. J.C. Bose, On the change of conductivity of metallic particles under cyclic electromotive variation. The Electrician **47**, 830–877 (1901)
62. S. Stathopoulos, A. Khiat, M. Trapatseli, S. Cortese, A. Serb, I. Valov, T. Prodromakis, Multibit memory operation of metal-oxide bi-layer memristors. Sci Rep **7**, 17532 (2017)

Magnetron Modes and the Chimera State

Victor J. Law and Denis P. Dowling

1 Introduction

The aim of this review paper is twofold. Firstly, bring real-world magnetron mode experiments and theory to the attention of the chaos community. This paper aims to achieve this by reviewing the development of the magnetron family (i.e. split-anode, cavity magnetron, rising-sun magnetron and relativistic magnetron) in terms of original work (therefore the most valuable documents, including patent applications). Secondly is to identify when and where synchronous and asynchronous modes and mode completion is present. It is hoped that this review will encourage chaos research into the magnetron family and the use of magnetron mode theory in natural coupled resonator systems.

By way of background electronic valve coupled ring oscillators that utilise multiple metal enclosures (high Q-factor resonators) in the form of the cavity-magnetron [1, 2] and the klystron [3, 4] were first developed in the late 1930s to replace the electrical (and patent blocking) limitations of the triode valve. Improvement in the cavity-magnetron in the form the coaxial cavity-magnetron [5] and the relativistic cavity-magnetron [6] came in the mid 1950s and 1970s, respectively. More recently the combination of quantum and microwave theory have produced compact microwave ring oscillators for consideration for stabilising atomic frequency standards intended for satellite navigation systems [7, 8]. In the development of all these oscillators, the reduction in mode competition (inherent output frequency instability) has been a major technical problem that had to be overcome. These instabilities where first described as having chaotic behaviour when phase locking two relativistic cavity-magnetrons was first reported using a length of waveguide [6]. More recently natural and real-world chemical and biological coupled ring resonator systems have been

V. J. Law (✉) · D. P. Dowling
School of Mechanical and Materials Engineering, University College Dublin, Belfield, Dublin 4 D04 V1W8, Ireland
e-mail: viclaw66@gmail.com

C. H. Skiadas (ed.), *Fractional Dynamics, Anomalous Transport and Plasma Science*, https://doi.org/10.1007/978-3-030-04483-1_2

studied using chaotic theory with particular focus on chimera states that contain the coexistence of synchronous (coherent) and asynchronous (incoherent) states in coupled ring oscillators [9, 10]. Given the long established mathematical constructs that describe modes within the magnetron family it is reasonable to consider that modes and chimera states have a converging origin, although they come from different scientific fields of study. This view point has been exemplified based on two papers presented at the 10th CHAOS2017 International Conference, Barcelona, Spain. The two contrasting papers are: *Electronic valve instabilities and modes* [11] and *Chimera: do they exist in the real world* where the second paper highlighted a presentation on coupled oscillators networks [12].

To address the objective of reviewing the development of the magnetron family it is first useful to follow the chronological order of events leading to their development. This may be found in the many academic peer-reviewed papers, books, conferences proceedings and patents. However, when it comes to a concise understanding of the driving force that lies behind the reduction, or even the eradication, of mode competition there is a porosity of record. In the case of patents the requirement is to lay out the innovation for all to see and how the patent relates to prior art. In this area Leconte was one of the first to survey more than 2000 magnetron patents application between the period of 1920 and 1942. Although mode competition was outside the scope of the work, the survey did successfully reveal both the historical evolution of the magnetron and the economic war pursued by rival international broadcasting industries involved in the manufacture of magnetrons [13]. For this reason US patent office publications are used alongside the academic publication resources. Using this approach it is hoped to eliminate the cloak of secrecy that surrounds the magnetron development during the Second World War (WWII) period. For example, due to military imperative (secrecy) the US patent office records reveal that Boot and Raddle's US patent was published in February 1951, some years 10 after the first US filing date in August 1941 [2].

The challenge in writing such a review is that word conventions change with time and place and scientific field of endeavour. For example; the word convention used to describe mode production and competition has changed with time, that is in the early years words favourable and unfavourable are prevalent in the literature and then moved on to support and unsupported as the mechanism of mode production and competition became clear. The cavity-magnetron has also been historically associated with the acronym RADAR (Radio Detection And Ranging) which was coined in 1940 by the U.S. Navy for public reference [14]. The term radar has since entered English and other languages as a common noun. Prior to this the terminology used to describe for radar was varied; such as 'object detection and location' in Europe and 'searchlight' in Japan: both of which are in text when referring to development in Europe and Japan. For more information regarding the terminology used see papers within: *The first international workshop on Cross-field devices* [15] and *CAVMAG 2010* [16].

The paper is structured as follows. Section 2 sets out the need for a high frequency oscillator, Sect. 3 follows the development of the magnetron family through to the cavity-magnetron with particular focus on lumped element equivalent electrical model (lumped element-EEM), mode component, mode degeneracy and mode jumping. Section 4 describes the international political and military involvement that governed the international development of the magnetron. Section 5 examines the post WWII era development of the magnetron. Section 6 looks at cavity-magnetron spectral noise. Finally Sect. 7 provides an overall paper summary.

2 Why Build a Magnetron

It can be argued that the magnetron has its origins in de Forest's 1907 audion tube patent [17] that was bitterly contested by Fleming who developed the thermionic diode valve [18], the patent [19] of which was subsequently bought by Marconi. The issuing litigation lasted some 10 years when it was compounded by Armstrong's work on de Forests audion tube (which later becomes known as the triode electronic vacuum tube) [20]. From an academic point of view reference [20] was republished in 1997 where both de Forest and Armstrong exchanged correspondence with the editor who had the final say [21]. By the mid1920s Appleton showed that the triode was capable of producing sufficient power levels at short wireless wavelength useful for investigating the ionosphere, for which he received the Noble prise in 1947 [22].

To prevent potential litigation problems, Governments and industrial companies around the world began research programs to overcome de Forest's patent. The initial and logical outcome was not to attempt to make an amplifier, but simply to replace the convoluted problem of grid electrode with a magnetic field to control current flow with the aim to produce space-charge [23] oscillations travelling through the $\boldsymbol{E}\times\boldsymbol{B}$ field, where $\boldsymbol{B}$ is the flux density of the magnetic field. The following paragraphs provide a summary of the development of magnetron family and the modes of oscillation encountered along the way.

3 The Magnetron Family

3.1 *Retarding-Field Triode*

The German scientists Barkhausen and Kurz were the first to utilise the electron transit-time effect, in a triode valve, to velocity modulation a signal with the grid at a positive potential relative to both the cathode and the anode [24]. Under these conditions electrons emitted from the cathode are accelerated towards the positive grid, where most pass between the grid wires and approach the anode where they are retarded back to grid and cathode. This electron dance continues back and forth through the grid until one by one they strike the grid wires. Hence the valve became called the "retarding-field triode", or "positive-grid oscillator". They found that their simple oscillator could operate in the frequency region of 1.7 GHz (17.5 cm) but with little useful power output. However, finding a constructive use of the three characteristic effects of the retarding-field triode namely; velocity modulation, bunching and power transfer from the beam to the circuit would be found later in the magnetron and klystron valve.

3.2 Diode Magnetron

The Swiss-German physicist Heinrich Greinacher was one of the first to use a cylindrically anode coaxially aligned to an inner cathode with a magnetic field superimposed parallel to the electrodes by a coil outside the glass envelope. Due to poor vacuum the attempt was only partially successful, but he did provided the first basic concept of electron precession within the $\boldsymbol{E} \times \boldsymbol{B}$ field. In 1921 Hull at the General Electric Company published his work on his coaxial diode valve [25]; a cross-sectional schematic of his magnetron (without the external coil) is shown in Fig. 1a.

Hull demonstrated that the strength of the superimposed magnetic field acts as a relay on current flow through the valve by restricting the electrons reaching the anode beyond a critical magnetic field strength. The mode of oscillation operation is one determined by the electron-transit-time between the cathode and anode. When operating just below the critical magnetic flux density, powers of 8 kW at 30 kHz were achieved. This regeneration frequency (f_r) follows the relation f_r = constant/B, where B is the correct flux density of the magnetic field to enable electrons to arrive at the anode and the constant is related to the electron transit-time between the cathode and anode and back. Hull called his tube the magnetron: the word is the synthesis of the words magnet and electron thereby the magnetron became part of the kenotron family of valves. By 1924, the Czech physicist Žáček [26] developed a magnetron with a solid cylindrical anode that generated frequencies up to 1 GHz (3 cm).

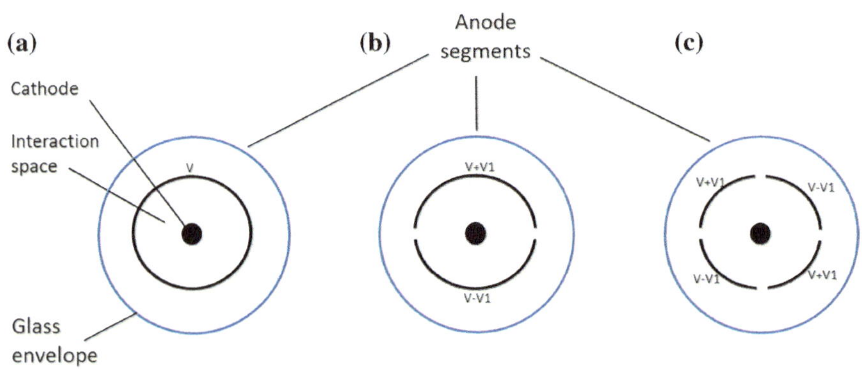

Fig. 1 Cross-section schematic of the anode segmentation development. Hull's single anode (**a**), 2 segment anode (**b**), 4 segment anode (**c**). The external coils are not shown for clarity

3.3 Split-Anode Magnetron

In 1924 Habann [27] produced the first split-anode magnetron. His valve had a central cathode wire surrounded by two plane or semi-cylindrical anodes, within an evacuated glass envelope, see Fig. 1b. Again a superimposed magnetic field parallel to the cylindrical arrangement was provided by an external coil. Using this arrangement a regenerative action of 100 MHz was obtained.

After graduate study in Germany, England, and America, Japan's best-known radio researcher in the 1920s–30s was Professor H. Yagi at Tohoku University. During this period (1927–29) one of his early doctoral students K. Okabe made a breakthrough into the centimetre wavelength using 2 and 4-segment split-anode design (Fig. 1b, c) [28]. In both cases the principle mode of operation was found to be related to the electron-transit-time and impulse or amplified negative resistance in which the frequency is equal to the natural frequency of the circuit.

A few years later, Posthumus famously derived a working mathematical formula that related the f_r to an even number of segments pairs, n, where $n =$ modal value of 1, or valves $=$ 2, 4, 6, so setting the scene for future development [29]. Even though the space-charge effect is neglected the approximate expression (1) shows this relationship,

$$f_r \approx \frac{4\pi n V_a}{r_a^2 B} \tag{1}$$

where V_a and r_a are the anode voltage and radius, also. Therefore for given constant V_a/B, increasing the number of segment pairs also increases f_r: which is in-line with experimental observation. The approximation term in Eq. (1) is used as magnetrons do not generally exhibit a zero-bandwidth power spectrum [30] but contains a phase noise component of approximately 100 s MHz centred on f_r. This is particularly true of the split-anode design where the anode segments are punched out with precision but with poor dimensional imprecision when fabricated together to form the composite split anode.

Crucially the opposing anode segments need to be operated at different RF potentials (V + V1 and V − V1, respectively, see Fig. 1c) so that the weak electric field surrounding the gaps can deflect the electrons as they sweep past these openings they induce a high-frequency resonant. This is because, as the electrons transit the gap some will be retarded and spiral out towards the most negative potential anode segment, while those electrons that gain energy are accelerated towards the cathode. In other words electrons reaching the anode regenerate the oscillation, but electrons retuning to the cathode do not. Moreover the retuning elections must move though the space-charge region, thereby temporarily disrupting the space-charge current before bombarding the tungsten-cathode and heating it further thus contributing to the split-anode magnetron's spectral bandwidth output.

The split-anode electron precession process is well exemplified in Gill and Britton's 1936 paper: *the action of a split-anode magnetron* [31] and the follow up

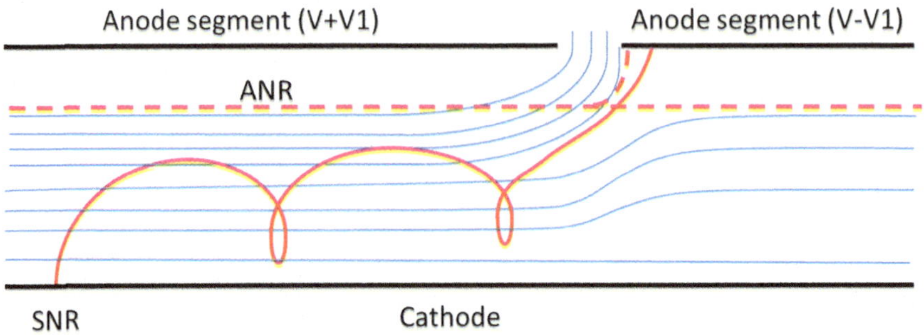

Fig. 2 A reconstruction of Heller's simplified scheme of two flat anode segments and cathode. The fine blue line represent the equipotential field; the solid red line is the path of single SNR electron and; the dash red line is the path of the ANR electrons

discussion [32]. In these two papers the contributors (Messrs. Awender, Tombs, Megaw, Gill and Britton) discussed alternative theories that may account for the static negative resistance effect and an amplified negative resistance which involves a resonance; where the oscillation period is equal to the time of precession from the neighbourhood of one anode gap to the next.

To differentiate between static negative resistance (SNR) and the amplified negative resistance (ANR) effect, Heller's flat plate split-anode picture may be used, which introduces no changes in the principle of their action [33]. Heller's picture is reconstructed in Fig. 2. Here the left-hand anode has a higher potential then right-hand anode, as may be seen from the course of the equipotential lines. Its potential is not however so great that an electron leaving the cathode with zero velocity could reach the plate. When an electron approaches the gap it enters a region where the equipotential lines are bent and lie closer to the anode. The electron will follow this course, and therefore reaches the anode with lower potential although it began its journey under the anode with a higher potential.

Electrons that just fail to reach the anode segment proceed to the next gap where they receive a second impulse thus gaining a second chance to reach the lower potential segment. This process repeats until the electron finally arrives at the segment with the lower potential. This repeating process is essentially the resonance, or amplified, type of negative resistance.

Hans Hollmann (Telefunken, Germany) developed the split-anode design further and filed an associated patent clam in 1936. The claim pertained to the rotational space-charge mode that operates at a higher frequency, but low efficiency, when compared to the amplified negative resistance mode. The claim (see Fig. 3) employs a plurality of sheet metal anode segments connected together using a further set of sheets metal straps of equal length to the segment length and where each straps forms a longitude cavity with a crescent shaped cross-section aligned along the segment gaps. The US patent claim was issued in 1938 [34]. Even though the anode was of a composite construction using welding or rivet technique and was still within an

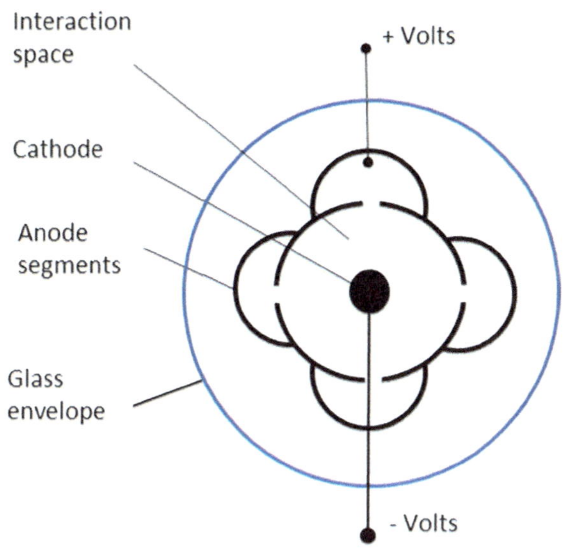

Fig. 3 Cross-sectional view of Hollmann's composite anode magnetron. Magnetic field removed for clarity

evacuated glass envelope that included the possibility of water cooling the design does foresee the multiple-cavity magnetron.

By the outbreak of WWII split-anode magnetrons where in full production in one form or another for shortwave radio use and in decimetre wavelength detection and location of aircraft systems that include supplementary visual-observation posts, such as the Home Chain defence system in Britain.

3.4 Block Anode Magnetron

In this section the transition from the split-anode to a block-anode magnetron is described. These magnetrons form a gradual transition in design thinking from one of a composite anode structure to the anode that is constructed using a single metal anode that has a plurality of cavities whistle still surrounded by glass vacuum envelope.

In the USA, Samuel at Bell Telephone Laboratories Inc., was the first (1934) to patent (US. 2,063,342) a magnetron containing 2 and 4 cavities machined out of a single anode block which was contained within a glass envelope with an external magnetic field applied [35]. The patent provides 5 schematics of 2 and 4 cavities surrounding a central cathode. A detail description of the electrical model of the components is given, however, no details of frequency and power measurements where given. Megaw later noted in his 1946 paper the electron discharge device lacked a satisfactory method of coupling the resonators to the load [36]. Moreover, there is little evident of Bell Telephone Laboratories developing their in-house knowledge. Indeed the company published 8 disclaimers to their original 18 claims (Official

Gazette March 8, 1938) and then took a different technical direction when they patented the silicon transistor amplifier in 1948 [37].

In 1928 Ukrainian institute of Physics and Technology (UIPT) was established at Kharkov State University (as part of the Soviet Union). In do so bringing together the Kharkov radio-physics community who become responsible for early investigation of the split-anode, a hollow anode with a glass envelope that allowed water cooling of the anode to achieve output powers of 5–7 kW at a wavelength of 80 cm. When the Laboratory of Electromagnetic Oscillations laboratory (LEMO) combined with UIPT to form UIPT-LEMO most of their work was published in German, so little of the work was reviewed in America, Japan or the UK. The *Zenit* three-coordinate (elevation, azimuth, and distance) aircraft detection and aiming system was a development that went un-noticed [38] and went on to be deployed around Moscow for air defence in 1941. However with UIPT-LEMO relocation to central Asia meant that the prototype *Zenit* system was not put into production.

It is also worth noting that the split-anode magnetron became of age 1935 when Henri Gutton's 4-segment anode was fitted with a 16 cm obstacle detector system on the ocean liner Normandie [39, 40].

By 1939, Tsuneo Ito at Tokoku University developed the 8 segment split-anode Tachibana (Mandarin-orange flower) which had a metal base with integrated water cooling within a glass envelope [41–43]. Later, with Yoji Ito, Tsuneo developed phase opposition, or push-pull, circuit that strapped alternate segments, thus in part reducing the phase noise. The Tachibana magnetron was further developed as the type M-3 magnetron that was contained within a glass envelope using a copper block anode that had radiating cuts giving rise to the name of *Kosumosu* (cosmos), or 'rising-sun'. The M-3 was able to operate at approximately 10 cm with a CW power of 500 watts and predating Boot and Randell's prototype magnetron by some months, although at a greatly reduced output power. In 1941 the M-3 was productionised to the M-312 magnetron, where the anode 'rising-sun' design employed vanes with cavities of uniform size and operated in pulse mode at 9.9 cm with a peak power of 6.6 kW [43].

In 1938 the technique of phase opposition, or push-pull was patented in the US by Frizz [44] where he describes how alternate segments are inter-connected by means of clips or connecting conductors plus the addition of an auxiliary electrode inserted at the cathode is used to control the radio frequency current. This was followed in April 1938 when Gutton and Berline filed a US patent where 4 claims are made relating to phase-opposition using a double 4-split anode where each split anode is mounted opposite to each other in such a way that their segments form a interdigital structure. The patent was issued in February 1939 [45] and was similar to the Japanese Corn-flower and lily type split-anode [41–43].

3.5 *The Leningrad Cavity Magnetron*

The mid to late1930s saw an interest in magnetron research with the aim to build robust devices that worked in the centimetre range with output powers of hundreds watts CW. Alekseev and Malairov at the *Nauchno-issledovatel institut-9* in Leningrad, U.S.S.R was the first to published water cooled solid copper anode with 2 and 4-cavities, without a glass envelope, that could reach these power levels [1]. Their work stopped after this publication, possibly due to the Stalinists purges at the time, clearly a case of published and be dammed [46]. Their paper was later translated from Russian to English by Bensen and republished in 1944 [47].

3.6 *The Birmingham Cavity Magnetron*

When Boot and Randall arrived at Birmingham University to work under the directions of the Australian physicist Mark Oliphant, their initial task was to miniaturise the Barkhausen-Kurz oscillator and to excite cavities by gas discharge. From their 1976 paper [48] Boot and Randall clearly state they were aware of the Varian brother's klystron [49] and Hansen and Richtmyer's mathematical treatment of the resonator enclosures made suitable for the klystron [3]. We may reasonable assume they were also aware of Hansen's early 1938 paper on the mathematical treatment of hohlraum type (spherical, cylindrical and prism) conducting enclosures [4]. Given the importance of these high Q-factor (30,000–50,000) conducting resonator enclosure (where Q is defend as π times the ratio of stored energy in the enclosure to the energy loss per cycle) there can be little doubt they applied this knowledge to the split-anode magnetron design that produced vanishingly small powers below 10 cm that were being constructed in: Denmark, France, Germany, Japan and the USA. Whether they knew of the Leningrad magnetron [1] is unclear. Through their reading, the Birmingham team were aware of the electromagnetic properties of the Hertzian wire dipole which resonates at electromagnetic wavelength of $\lambda = 7.94d$, where d is the length of the loop [1, 50].

With the publication timing of the resonator enclosure papers [3, 4, 49] and the implications of the Hertzian loop it is not surprising that Boot and Randell came up with a 3-dimensional version of the Hertzian dipole that forms the re-entrant cavity design, were the physical size of each resonator is fixed by the desired output frequency. By May 1940, and some 3 months after of Hansen's US patent (*High efficiency resonant circuit*) [51], the Birmingham team made their 6-slot-hole cavity magnetron: it was constructed with 6 holes of diameter, $D_{cav} = 1.2$ cm, within a copper block of height, $h = 4$ cm. Interestedly, the holes diameter and circumference spacing were a determined by a Colt revolver barrel that was used as a drilling jig. The slots (width, $w = 1$ mm and thickness, $t = 1$ mm) where cut radially from the anode inner diameter ($D_a = 1.2$ cm) to each hole with the annular spacing between the anode and cathode defining the interaction spacing. Hence each slotted hole

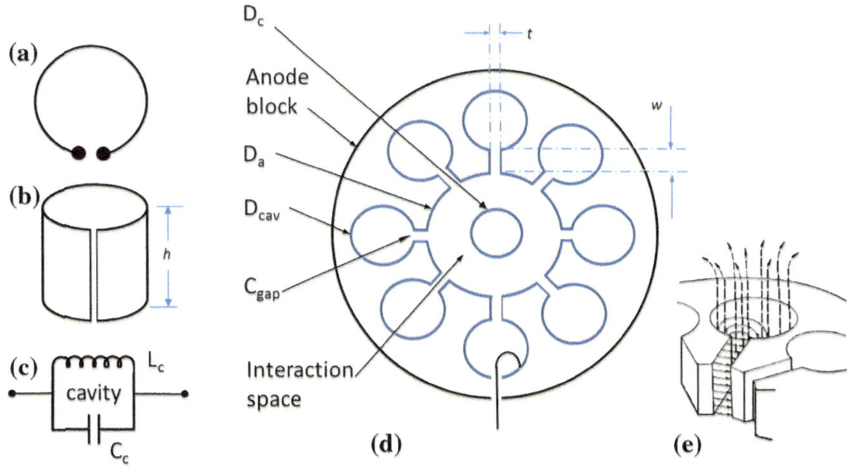

Fig. 4 Magnetron re-entrant cavity development: **a** Hertzian loop, **b** 3-dimensional Herstzian loop, **c** equivalent LC lump EEM, **d** cross-sectional view of a N = 8 cavity magnetron and, **e** electromagnetic field-lines within a cavity

approximated to an electromagnetic wavelength of 1.2 cm × 7.94 = 9.5 cm. See Fig. 4a, b, c, d, e. The 4 cm height of the anode block, and hence cavity height, was selected to fit the electromagnet gap size available at the time and was later found not to be critical to the magnetron output frequency.

The French influence came from Maurice Ponte at the Compagine Génerale de Télégrahpie Sans Fil who escaped to Briain before the fall of France. Ponte provided the British team with the French M-16 magnetron that had a heated oxide-coated cathode and 8-slot-hole cavity enclosed within an evacuated glass envelope giving peak power of ~1 kW at 16 cm [36, 48, 52].

Boot and Randall believed that the oscillations within each cavity circuit were strongly coupled together through the combined action of the axial magnetic field and radial electric field that forced the electrons to rotate past the re-entrant gaps. The effect being to produce a synchronised feedback electromagnetic slow-wave structure: that is an electromagnetic wave that propagates at a velocity much smaller when compared to that of light. Coupling the radio frequency power out of the interaction space was provided by means of an output coupling probe, see Fig. 4d.

The removal of external tuning circuits in the Boot and Randell design is more than compensated when you consider that a few watts of almost spurious radiation was turned into 100 kW of well-defined signal generated by a compact practical device. The cavity magnetron advantages can be listed as: (a) minimal copper loss design of the single machined anode block that incorporate all the cavities provided a less complex manufacturing process that reduces segment miss-alignment and an improved anode surface smoothness, both of which untimely produce the high Q-factor requirement for high energy storage within the cavities. And (b), placing the vacuum inside the anode structure removes the need for a glass envelope. These advances over a short period time produced an all-nearly metal design with efficient cooling of the anode block that allowed an increase output frequency as compared to previously patented magnetron designs. Moreover the light-weight design enabled centimetre radar sets to be made airborne and seaborne thus fulfilling Watson-Watt and Wilkins memorandum: *The detection and location of aircraft by radio methods* [53].

3.7 Lumped Element-EEM of Cavity Magnetron

Although monolithic structures of the anode block cannot be strictly modelled using the lumped element-EEM, a rough estimation of the resonant frequency (f_r) of the cavities may be performed. Indeed three patents [35, 51, 54] and three journal papers [3, 4, 49] stress that those skilled in the art of EEMs can use the lumped element [inductance (L) and capacitance (C)] approach with a high degree of success. This paper takes the same approach and where necessary use electromagnetic field theory to estimate structural induced inductive and capacitive fringing effects. In addition the EEM is tested against the Leningrad, Birmingham and GEC cavity magnetrons.

First consider the conventional parallel resonant circuit (Fig. 4c) whose basic resonant frequency (f) is given by the first order Eq. 2.

$$f = \frac{1}{2\pi}\sqrt{\frac{1}{LC}} \tag{2}$$

where L is the inductance value of the individual cavity and C is the capacitance value between each re-entrant gap. When attempting to generate an EEM of the cavity magnetron operating in the desired π-mode, the individual inductors and capacitors are all connected in parallel, therefore the effective inductance is L/N and effective capacitance is NC, where N is the number of resonators. Equation 2 may now be written using the expanded forms of L and C to produce the approximate expression (3).

$$f_r \sim \frac{1}{2\pi r}\sqrt{\frac{Nt}{\varepsilon_0 \mu_0 \pi w}} \tag{3}$$

where f_r is the resonant frequency of the system, N is the number of re-entrant gaps, t and w are the thickness and width of each re-entrant gap and $\varepsilon_0 = 1.25 \times 10^{-6}$ F/m and $\mu_0 = 8.85 \times 10^{-12}$ H/m. Note, as the height of the cavity (h), see Fig. 7b, appears in the numerator and denominator in the L and C formulas, these dimensions are cancelled out making approximate expression (3) independent of h thereby supporting the observation that cavity height is not critical to the magnetrons resonant frequency.

Table 1 Structural anode geometry of the Leningrad, Birmingham and the GEC E1189 (# 2 and # 12) cavity magnetrons

Magnetron	D_{cav} (cm)	D_a (cm)	h (cm)	$t \times w$ (mm)	f_r/λ (GHz/cm)
Leningrad 4 cavity	0.9	0.6	2	1 × 1	~3.3/9
6 cavity Birmingham	1.2	1.2	4	1 × 1	~3/9.8
6 cavity E1189 # 2	1.2	1.2	2	2 × 2	~3/9.8
8 cavity E1189 # 12	1.2	1.6	2	2 × 2	~3/9.8

A second term that contains the capacitance fringing correction (a) may, or not, be aided depending on the fringing rejection criteria d $\ll \sqrt{}$A holds true. In the case of the cavity magnetron where the re-entrant gap is formed by two opposing surfaces, with t and w being of similar dimension the criteria is always negative: thus the term is employed and its value varies between 0.53 and 0.73 depending on the cavity geometry, see in Eq. (4). In the case of the cavity magnetron re-entrant configuration the lefts-hand-side term in the in Eq. (4) is used, thus the fringing capacitance contributes about 50% [7, 8].

$$\sqrt{\frac{1}{1+2.5\frac{t}{w}}} \leq a \leq \sqrt{\frac{1}{1+\frac{t}{w}}} \tag{4}$$

The fringing inductance of the slotted cavity may be treated in a similar way. However in this work the value of the fringing inductance (b) is extrapolated from resonant frequency and anode geometry measures of the magnetrons. The procedure is performed by introducing a total fringing correction coefficients $\boldsymbol{\alpha} = (a\ .\ b)$ and inserting this term outside of the square root sign of expression (3), see approximate expression (5).

$$f_r \sim \frac{\alpha}{2\pi r}\sqrt{\frac{Nt}{\varepsilon_0 \mu_0 \pi w}} \tag{5}$$

The approximate expression (5) may now be used in meaningful way when comparing the anode geometry of the Leningrad, Birmingham University and GEC prototypes magnetrons. Table 1 provides a structural comparison of their anodes and measured f_r values. The first feature of note is their remarkable similarity in anode structural design and the measured operating resonant frequency. Here again the valve of h (within the limits 20–40 mm) does not appear to influence f_r.

Applying the approximate expression (3) to the four magnetrons measured values of fr and their associated anode geometry data an estimation of $\boldsymbol{\alpha}$ is now computed. In all cases, to 2 decimal places, a value $\boldsymbol{\alpha}$ is found to be 0.27, see Table 2, column 4.

Table 2 Comparison of measured and estimated f_r for the Leningrad, Birmingham and two GEC cavity magnetrons (E1189 # 2 and # 12)

Magnetron	f_r/λ (GHz/cm)	Equation 3 (GHz)	Coeff α	Coeff a	Coeff b
Leningrad 4 cavity	~3.3/9	11.9	0.27	0.53	0.51
6 cavity Birmingham	~3/9.8	11	0.27	0.53	0.51
6 cavity E1189 # 2	~3/9.8	11	0.27	0.53	0.51
8 cavity E1189 # 12	~3/9.8	11	0.27	0.53	0.51

The computed value of α is given in column 4

Given this and the estimation of coefficient a = 0.53 used in the extrapolated fringing inductance valve is equal to b = 0.51 in all cases, see Table 2, column 6.

3.8 Mode

It was found that the cavity magnetron's output changed from pulse to pulse, both in frequency and phases due to mode jumping. The mode also varied from one magnetron to another. Unfortunately, the difference in frequencies of the modes, or mode separation as it is called, is not sufficient to stabilise the magnetron power output. The term moding arises from mode competition from the mode closest in frequency to the π-mode. In 1941 with or without the knowledge of the phase-opposition technique [43] James Sayers took an interest in these mode jumps. He found that due to the multiplicity of resonators used a number of modes are possible were each mode is characterized by the mutual phase displacements between two successive cavities. Fortunately the number of possible modes is limited by the fact that the resonator system (N-cavity resonators and the interaction space) is a closed system. Thus the resonant modes of an anode block containing an even number of N-cavities is characterized either by an integer n, defined as the total electrical phase change around the anode block measured in revolutions (N/2); or by the phase difference ($\Delta\phi$) between successive cavities. For an anode block with N cavities these two quantities ($\Delta\phi$ and n) are related by the Eq. (6).

$$\Delta\phi_n = \frac{2\pi n}{N} \tag{6}$$

For an $N = 8$ cavity anode block, $n = 0, 1, 2, 4$; where $n = 0$ is not used as its electric fields do not interact with individual cavities, and thus does not determine the mode frequency. In the special case mode $n = 4$, successive segments are in antiphase (that is they have equal and opposite potential, $\Delta\phi = \pi$). This mode is automatically

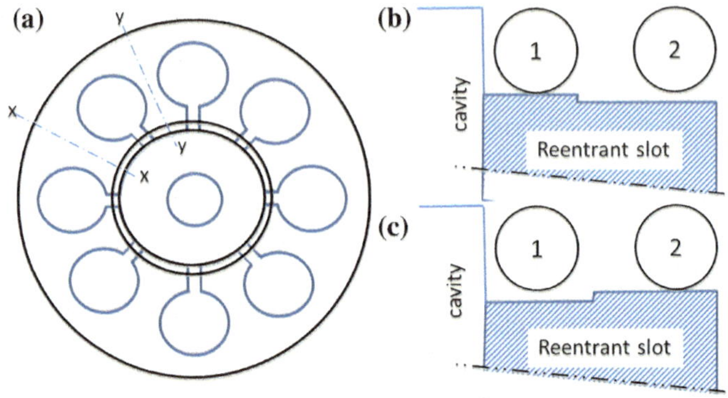

Fig. 5 Cross-sectional schematic views of a double strapped N = 8 cavity magnetron: **a** the double strap system, **b** sectional view of xx showing outer strap (1) connected to segment and inner strap (2) not connected: **c** sectional view of yy showing outer strap (1) not connected to segment and inner strap (2) connected

selected as it does not suffer from degeneracy thus the mode frequency determines the electromagnetic field pattern uniquely. It also follows that for the mode $n = 4$, one spoke revolution around anode block equates to N/2 RF oscillations. To provide sufficient separation and hence selection and stabilization between the neighbouring modes ($N/2 \pm 1$) Sayers developed the double ring strapping system where the first ring connects the even numbered segments and the second ring connects the odd numbered segments (Fig. 5a) so fixing the phase difference between the segments to a value of π (180°). This procedure of making two coupled oscillator systems that are connected through the phase delay of the slow-wave structure within the interaction space is now called "π strapping".

An additional feature of the double π-strap is that the wavelength of the π-mode is increased without significantly affecting the other modes. It thought that the effect arises through the following manner. In the mid-plane of each strapped segment the magnitude of the voltage and phase is the same as the strap whereas for the unstrapped segment parallel capacitance is added to the capacitance of the re-entrant gap thus increasing the wavelength of the system, see cross-sectional view xx in Fig. 5b. For the other modes, the individual strap capacitances are not all in phase with one another and therefore do not contribute so much to the system capacitance. Figure 5c shows the reversal of strap connection at cross-section yy.

Now consider the initial magnetron start-up period which last approximately 3 s. Firstly under a constant magnetic field parallel to the axis of the cathode the radial electric field from the applied dc potential grows to produce an increasing $\boldsymbol{E} \times \boldsymbol{B}$ field. In this time period emitted electrons from the cathode move away from the cathode surface to become part of a space-charge within the interaction space thus the spatial-temporal arrival of the electron at the anode segments alters greatly. Under these conditions mode competition between both ring oscillator systems are generated and destroyed until the stable π-mode spoked wheel is formed (3 spokes for 6 cavities and 4 spokes for 8 cavities) with the electrons densely packed (bunched) at the interface between the tips of the spokes and the anode segments. At re-entrant gaps the electrons are slowed down and pass their energy on to the radio frequency field leading to the next anode segment becoming loaded a little more negatively, so producing a rotation of the spoked wheel. However, unlike the split-anode design, the phase focusing effect enables the transferred energy to be passed on to the cavities and enforce the cavity microwave oscillations This synchronisation of the tangential speed of the space-charge spokes with the phase velocity of the slow-wave structure is generally known as the Buneman–Hartree resonance condition.

3.8.1 Mode Components

Douglas R. Hartree's contributed significantly to the theory of atomic structure and the operation of the magnetron. However, in the case of the magnetron mode of operation he did not publish in journals, but did write extensive technical reports for the British government during WWII [55]. To some degree this has obscured his contribution. Here we lay out his contribution in terms of Hartree voltage and the Hartree component, or harmonic. Firstly, Hartree was one of the first to consider what has become known as the Hartree voltage: the lowest voltage at which oscillation begin (provided at the same time that $\boldsymbol{B}$ is large), so that the undistorted space-charge does not extend to the anode. Secondly, mode selection and rejection ($n \neq N/2$), where he theorised that oscillations in a given mode takes place when electrons within the rotating space-charge couples to the slow-wave structure via a harmonic of one of the resonant frequencies within the closed slow-wave structure. The Hartree component or harmonic (γ) may be represented by Eq. (7).

$$\gamma = n + mN \tag{7}$$

where m is a whole number and, n and N have already been defined above. Although the number of components is large, only values of γ for $\pm\, m$ up to 2 are observed due to the drop-off in intensity of these components is proportional to (cathode radius/anode radius). The operating modes are designated using the nomenclature $\gamma/n/N$. It is convenient to visualise the electron configuration as γ spokes. Thus the π-mode in an 8-cavity magnetron, when excited through its fundamental, is represented by the symbol 4/4/8, see Fig. 6. In the case of the n = 3 mode when excited by its fundamental the designation is 3/3/8, and if the n = 3 mode is excited through the y = −5 component, the designation is 5/3/8. (*N.B. a modern interpretation and use of the Hartree harmonic may be found in resonant drive circuit for plasma jets* [56]).

Aligned with Hartree's harmonic is that the space-charge spokes contain strong shear forces that drive diocotron instabilities close the cathode in the magnetron start-up period. See the annotation (diocotron shear interface) in Fig. 6. The view of different radially spaced electron cloud layers having increasingly higher rotation

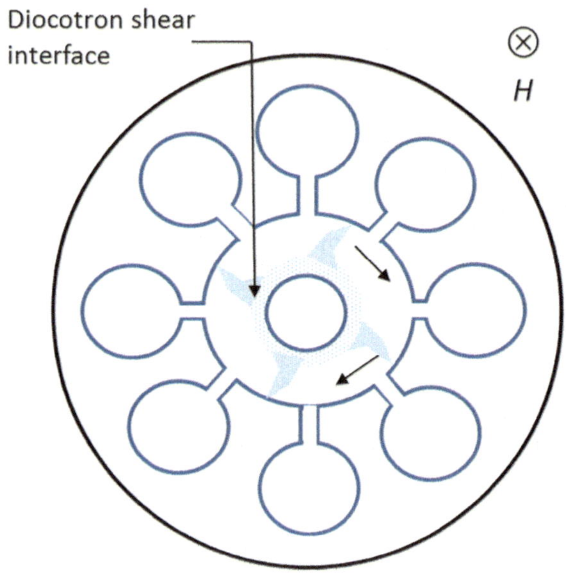

Fig. 6 Cross-sectional schematic view of N = 8 cavity magnetron with the space-charge wheel rotating about the cathode. The magnetic field is going into the page

velocities as the spokes reach the re-entrant cavities and therefore allowing the layers to pursue, or chase, each other has become known as the diocotron mode [57–59].

3.8.2 Mode Degeneracy

Within the circular symmetry of the non-strapped anode interaction space mode degeneracy occurs due to waves traveling in opposite direction. The modes are designated by their mode number N/2 ± 1, 2 etc. The non-degenerate N/2 π-mode and the 0-mode (at infinite wavelength) are not included in this class of mode. For example, the N/2 − 1 mode travels in the opposite direction to the N/2 + 1 mode but has the same non-sinusoidal radio frequency field pattern as the original travelling wave at any given moment in time. In reality the symmetry of the interaction space is perturbed by the output probe, and by imperfections in the machining of the anode block, so making these modes less favourable in the start-up period. The degenerate modes also haves a role in mode skip (where the magnetron does not fire on every pulse) and, mode shift (where the mode of operation changes during the pulse).

3.8.3 Mode Jumping

The requirement of adequate separation between the desired mode and any other interfering mode is fundamental to the magnetron operational design. From the above the descriptions it can be seen that the interaction space has an ensemble of modes that can contaminant and compete with each other. For the interaction space design to automatically select the desired π-mode and reject others it is desirable that there is equal bandwidth separation between the π-mode and interfering modes, in conjunction with a lower threshold voltage for the π-mode as compared to interfering modes: giving little or no chance of unwanted modes to compete with the π-mode.

The π-strapping techniques goes some way to remove mode skipping and jumping at the 10 cm wavelength, but the technique was found to be difficult to achieve in magnetrons operating at wavelengths below 3 cm. The strapping limitation is mainly due to difficulties in manufacture and the inductance of the straps. However, by representing the magnetron cavities by a lumped-element EEM it was found that the rising-sun anode structure, consisting of alternate deep and shallow cavities, led to adequate separation of π-mode frequency from other unwanted modes.

4 The Military Imperative

The invasion of Manchuria by Japan 1931 that was forerunner of the WWII meant that research into high frequency oscillators and their incorporation in the radio frequency equipment that could detect both aircraft and ships became the subject of many secret military projects; The information of which could not be divulged through patents or journal papers. It was only after secession of the pacific war that journal papers and magazines openly reported on the magnetron development and their contribution to radio searchlight and location systems. This section therefore relies heavy on the articles of: Megaw (1946, [36]), the report prepared by United State Army Air forces; survey of Japanese radar capabilities published in 1946, [60, 61]), Döring (1992, [62]), and Nakajima (1992, [42]). In view of this imperative, it is the aim of this section to provide a picture of military use and advances made during the period of 1936 to 1945.

4.1 Tizard Mission

Arguable the Tizard mission (British Technical and Scientific Mission to the United States of America and Chanda) changed the technological course of WWII in the late summer of 1940; the aim of the mission being to obtain industrial resources and exploit the military potential British research and development up to that point. Tizard's "brief case" therefore held the sealed-off magnetron type E1189, series N.12 developed by Megaw at GEC, enigma code breaking information and all what Britain knew about building an atomic bomb. As regards to the said magnetron, the American's scientists X-rayed it and found 8 cavities whereas the accompanying drawings stated 6 cavities thus giving rise to suspicions. Megaw was telegraphed and after a short time was able to remember that the first 10 magnetrons had 6 cavities and number 12 had 8 cavities, immediately the drawings were updated and a developing international incident was avoided. For this extraordinary mission, Britain would receive financial and industrial help for their war effort. By September, the Massachusetts Institute of Technology had set up a secret laboratory; by November, the cavity magnetron was in mass production; and by early 1941, portable airborne radar had been developed and fitted to both American and British planes with pulsed transmitter powers of ~100 kW at 3 cm in 1940 to ~2 MW at 10 cm by 1944. The radar sets could also be fitted with the American klystron amplifier as they would be more likely to be destroyed in the event of a plane crash [50].

To place the success of the Tizard mission into context, it was not until the night of the 2nd to 3rd of February 1943 that a high power 10 cm cavity magnetron and a reflex klystron fell out of the sky into Germany's hands. This came about when a Stirling bomber was shot down south east of Rotterdam with its secret ground scanning radar equipment virtually undamaged. Germany made an almost exact copy which they named the "Rotterdam Great" [62]. Inevitably similar occurrences happened in the pacific in 1945 when B29 bombers were shot down [60, 61]. Fortunately for the Allies the capture of the centimetre radar secret came too late to have any significant effect on the outcome of the war.

4.2 The Tripartite Pact

During the period between the two World Wars, wireless technology in Japanese universities, especially the Imperial (government-financed) universities, was on a par with that in Europe and the USA. Academic publications encouraged along with academic and industrial exchange to western universities were encouraged. The Japanese invasion of Manchuria in, 1931 and the 1940 Tripartite Pact with Nazi Germany and Fascist Italy naturally meant that wireless research and oscillators such as the magnetron become sensitive areas of military research. It is also fair to say that the top military leaders sincerely felt that the application of radio in detection and ranging was as defensive tool that did not suit their offense thinking. In addition when it came to transferring scientific research into military use problems arose due to rivalry and jealousy between the Imperial Army and Navy; though the Pact did encouraged strong links with universities and institutions.

By late 1940 technical missions to visit Germany and Italy were arranged with the aim to exchange developments in military technology and information. The visits to Germany resulted in Japan receiving the H2S ground scanning radar plans and details of a pulse-modulated microwave radio range finder set. The visit of the assistant naval attaché to Berlin to Italy, after the British aerial torpedo attack on the Italian Navy at anchor in the harbour of Taranto 1940 is likely to have influenced japans attack on Pearl Harbour. At the time Italy's naval and coastal radar capabilities relied on RCA and Philips triodes [63].

Japan also had early aircraft searchlight systems based on the continuous wave Doppler affect. This came about when in 1938 a team of engineers from the Research Office of the Nippon Electric Company (NEC) were forming high-frequency transmitter coverage tests. They found that rapid fading of their signal occurred whenever an aircraft passed over the line between the transmitter and receiver. The disadvantage, however, was that it did not give a definite location. Masatsugu Kobayashi, the team manager, recognized that this was due to the beat-frequency interference of the direct signal and the Doppler-shifted signal reflected from the aircraft and suggested to the Army Science Research Institute that this phenomenon might be the same that Yagi and Okabe had found earlier in 1936. Thus by 1940 the Imperial Army's Bi-static Doppler Interference Detector was placed in service. The system operated between 4.0 and 7.5 MHz and involved a screen of widely spaced stations that could detect the presence of an aircraft at distances up to approximate 500 km.

5 Post WWII Magnetron Development and Use

Towards the latter part of WWII magnetron research in the America, Britain and Japan turned to magnetrons that could generation millimetre wavelengths for greater target definition, higher microwave powers and non-radar use. These three research goals required the slot-hole cavity magnetron to be developed further. These developments are listed below.

5.1 *Cavity-Magnetron*

With regard to increasing the cavity magnetron turning range the 1949 US patent was given to Percy L. Spencer (Raytheon) [64]. This was followed a by two more US patents in 1951: Randell and Boot (English Electric Valve Company Limited) cavity-magnetron [2]; Sayers's (English Electric Valve Company Limited) double ring strapping system [65]. These were shortly followed by Okress and Reed (Westinghouse Electric Corporation) US patent relating to the echelon strapping system [66]. That modified the Sayers's strapping system in such a way to reduce high voltage arcing, hence power loss, between straps and non-directly connected segments, two further significant patents were published in 1952: Dodds (Radio Corporation of America) relating to magnetron cavity pre-turning within the manufacturing stage [67], and Flowers (Western Electric Company) automatic magnetron aging circuit [68].

5.2 *Broader Applications of Cavity-Magnetron*

By the late 1940s the Raytheon Company transferred their military magnetron technology to medical therapeutic purposes (diathermy) and commercial food preparation using their Radarange oven [69]. When the Raytheon basic patent expired in 1967 new low-cost table-top microwave ovens became available that contained the packaged magnetron (magnetron with integrated magnet). Later (1975) the first planar magnetron patent for sputtering of materials was issued to Corbani (Sloan Technology Corporation) [70] and, possibly the first electron cyclotron resonator plasma etcher patented was issued to Suzuki et al. at Hitachi in 1978 [71]. Further to these patents the first low-cost microwave plasma etcher patent was issued to Aaron Ribner in 1989 [72]: examples of use being plasma etching of contaminated ceramics [73]. These new use of the magnetron marked a radical change from launching kW microwave radiation to only measuring voltage fields of a few 10 s mV m^{-1} for radar purposes [53] to one of 400–900 W microwave heating and plasma treatment of materials.

5.3 Rising-Sun Cavity-Magnetron

The collection and handover of the Japan's magnetron and radar technology to the Allies at the end of WWII [60, 61] spurred-on USA researchers to investigate the vane-type magnetron, with the aim of improving both frequency range and output power without the need for strapping. Within three years a triplet of papers appeared in the Journal of Applied Physics. The results where explained by Millman and Nordsieck [74] using the lump element-EEM and an electromagnetic field theory model by [75, 76]. The three papers have proved to be important in our understanding of how the π-mode is automatically selected and other modes rejected. To aid the discussion, Fig. 7a shows a cross-section view of the 24 GHz (1.25 cm) 18-vane rising-sun cavity-magnetron in question. The word segment is dropped for the preferred term vane as this describes the configuration more accurately. In addition [75] referred to the Japanese type M312 rising-sun cavity-magnetron [41–43] as a possible design source of their alternating cavity shaped magnetron.

For this rising-sun configuration the ratio of the resonator depths of the two different cavities $r_i = 1.6–2.0$ was found to be entirely satisfactory for π-mode operation.

Figure 7b shows an elementary section of the lumped element-EEM of an 18-vane alternating cavity rising-sun anode block. In this figure the left hand $L_1 - C_1$ resonator represents the smaller cavity and the right hand side $L_2 - C_2$ resonator represents the larger cavity. Capacitive susceptance (C_s) to a common point (cathode) is also present. Each resonator also has an associated wavelength, λ_1 and λ_2 that are set by the resonant circuit L and C values, where r_1 is the ratio of the resonator depth of the two different cavities, and in the value of σ, the ratio of the cathode diameter to the anode diameter. To exemplify these conditions the following remarks are taken from Ref. [74, 75].

(a) *When r_1 is too small, competition from the $n = [(N/2) - 1]$ mode interferes with stable π-mode operation of the magnetrons.*
(b) *When r_1 is too large, mode competition from members of the long wavelength multiplet having $n \geq 3$ prevents satisfactory operation.*

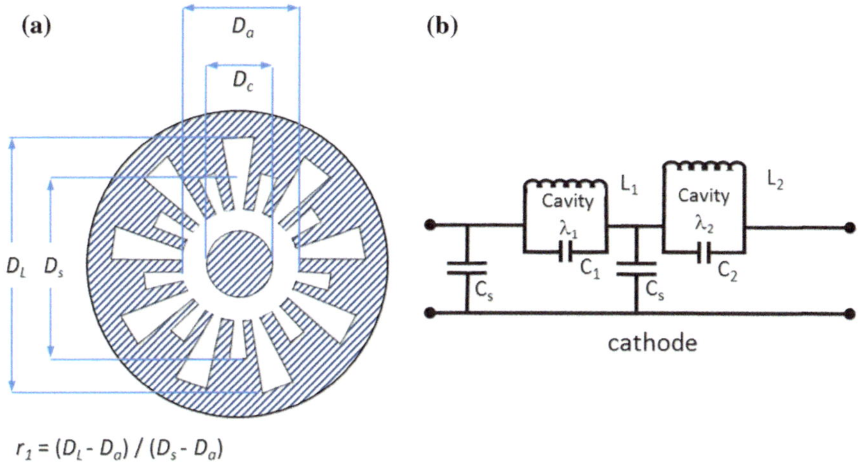

Fig. 7 Diagram of the 18-vane rising-sun cavity-magnetron: **a** cross-sectional view of the anode and cathode block: **b** elementary circuit (containing two cavities) used in the elementary LC lumped circuit. The circuit continues around and closes on itself using N/2 elements to complete the model

(c) *The range of satisfactory values of* r_1 *decreases as the number of resonators is increased. For magnetrons in the voltage range under discussion, satisfactory open-cavity vane type magnetrons with* $N > 26$ *can probably not be designed.*

Given these conditions, r_1 is usually set between 1.5 and 1.7. The following lump element-EEM assumptions can now be made and are. Firstly, if the values of C_s are small they can be neglected to give an approximate picture of the modes. Secondly, when the valves of $C_s = 0$, the circuit enables three resonant modes to be generated: these are when either of the two parallel impedances ($L_1 - C_1$, or $L_2 - C_2$) is at infinite, or when the two parallel impedances are opposite and equal. Thirdly, if a small amount of coupling between the resonators is introduced the first and second resonance are split into two closely spaced resonances each, while the third is not spit. These three resonant groups correspond to the low wavelength (high frequency) group n = 5, 6, 7, 8, the long wavelength (low frequency) group n = 1, 2, 3, 4 and the π-mode n = 9.

Figure 8 shows the computed mode spectrum for n = 1–9 of the 18-vane alternating cavity rising-sun anode. For comparative purpose the computed mode spectrum of an equivalent 18-vane double-strapped anode is also shown. In this figure the system free space wavelength is plotted on the vertical axis and n is plotted on the horizontal axis with all the data points are taken from ([74], Fig. 1b, c). Here it can be seen that the mode spectrum forms a number of groups equal to the number of different sizes of resonators used. In the case of the rising-sun anode two groups are formed due to the two different sizes of resonators used: where the large separation or gap in the mode spectrum is formed near n = ½ ($N/2 - 1$). Whereas for the 18-vane double

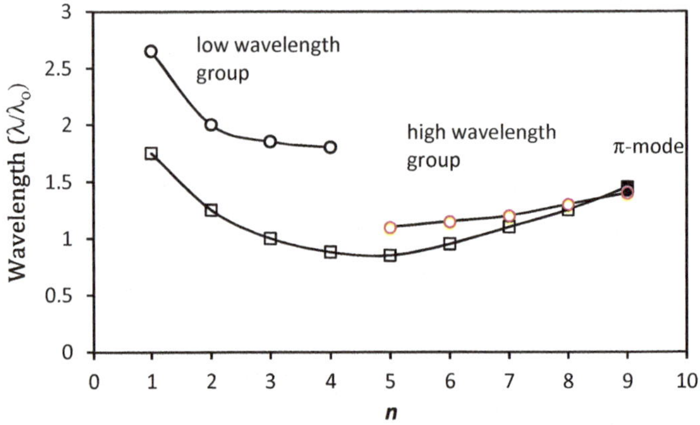

Fig. 8 Typical mode frequency spectrum associated with an 18-vane double-strapped anode (open squares) and the alternating cavity rising-sun anode (open circles). The black circle and square represents the π-mode in each case

strapped anode no large separation is presents as there is only one set of equally sized resonators.

To concluded this section it is noted that in Robert Moats Ph.D. thesis (Interaction of modes in a magnetron, oscillators, MIT, 1950 [77]) treated the rising-sun magnetron as triode oscillator with two degrees of freedom' that is, two modes in the resonant circuit. In doing so Moats based his work on B. van de Pol's non-linear theory (on oscillation hysteresis in a triode generator with two degrees of freedom) [78].

5.4 Coaxial Cavity-Magnetron

The removal of unwanted modes at high frequencies without strapping or alternating cavities was first realized by Collier and Feinstein in 1950s [5]. Their US 2,854,603 patent (1958) used a pair of coupled cavity resonators that have a common connecting wall. The outer torroidal coaxial cavity is designed as a high-*Q* resonator that oscillates in the TE_{011} and that is set at different frequency outside of the turning frequency range of the inner anode-cathode cavity resonator. Electromagnetic coupling between the two cavities systems is achieved through alternating pairs of anode vanes by slots extending through the common wall and where the anode vanes of the inner cavity are approximately one quarter-wave long in frequency range of the TE_{011} oscillations of the coaxial cavity. With the inner cavity operating in the π-mode every other cavity is in phase with the coaxial cavity (Fig. 9).

Because the outer torroidal coaxial cavity has a high *Q*-factor, electromagnetic coupling to the slow-wave structure within the interaction region stabilizes the π-

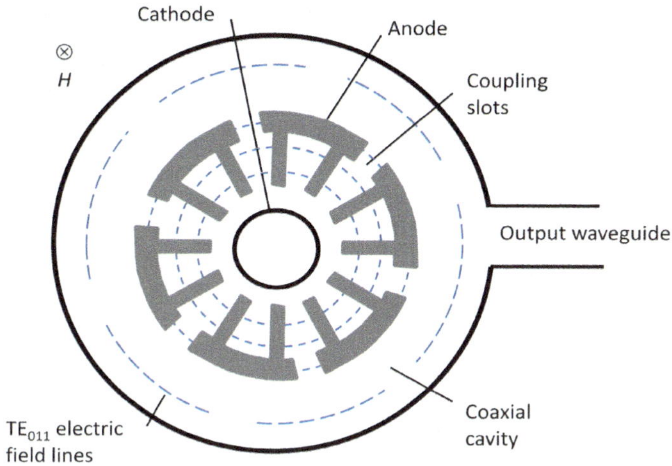

Fig. 9 Cross-sectional view of a 12-vane coaxial cavity-magnetron. The magnetic field is going into the page

mode and dampens all other modes. This action both increases the magnetron output frequency stability and reduces spectral noise when compared to the original cavity-magnetron. The design also allows rapid (millisecond) auxiliary tuning via movable ferrite [79] and dielectric strips and paddles [80] without introducing mechanical vibration. All of these factors make the coaxial cavity-magnetron the preferred choice for modern civil ground and airborne radar.

5.5 *Relativistic Cavity-Magnetron*

This section looks at the A6-type 6-vane relativistic magnetron where the term 'relativistic' refers to the use of high-voltage electron beams with velocities close to the speed light. Thus the fundamental difference between the relativistic magnetron and the cavity magnetron is the application of diode voltages >360 kV and explosive emission from the cathode by removing the current limitations connected with cathode emission and space charge. To achieve these conditions the current limiting π-strapping is removed and thermionic emission cathode replaced with a smooth wall field-emission cathode. The anode is formed as a slow-wave structure that contains 6 circumferentially spaced radially oriented cavities towards the cathode with output coupling probe replaced by a waveguide with its iris positioned in one of the cavities to enable radial extraction of the radiation energy out of the magnetron. With these changes high-energy density physics experiments can be realised. Again the interaction space is evacuated to better than of 10^{-7} Torr. In 1976 Bekefi and Orzechowaski published their relativistic magnetron that produced microwaves power of

Table 3 Diode aspect ratio (σ) and mode for the Leningrad, Birmingham and E1189 # 2 and # 12 plus the 6-vane A6 relativistic magnetron

Magnetron	σ	Mode	References
Leningrad 4 cavity	2	π	[1, 85]
6 cavity Birmingham	0.3	π	[35]
6 cavity E1189 # 2	0.3	π	[35]
8 cavity E1189 # 12	0.23	π	[35]
Relativistic magnetron	<1	π	[82]
Relativistic magnetron	2.98	π and 2π	[82, 83]
Relativistic magnetron	4.7	π, 2π and 5p/3	[83]

1.7 GW at 3 GHz in 30 a nanoseconds pulse [81], this was followed by their US patent published in 1980 [82].

The process of mode competition in relativistic magnetrons have been experimentally studied by Palevsky et al. [83] and simulated by Chen et al. [84] as a function of diode aspect ratio $\sigma = r_c/(r_a - r_c)$ by varying the cathode and anode radii r_c and r_a. Their studies revealed that magnetrons with large anode-cathode gaps (small aspect ratio, $\sigma < 1$) favour oscillation in the π-mode ($n = 3$, $\Delta\phi = \pi$) for a wide range of diode voltages and magnetic field values, whereas magnetrons with a small anode-cathode gap (large aspect ratio, $\sigma > 1$) tend to oscillate in the π-mode and 2π-mode ($n = 6$, $\Delta\phi = 2\pi$), with the 2π-mode performing better than the π-mode. Moreover at very small anode-cathode gaps ($\sigma = 4.7$) the $5\pi/3$-mode can be excited where the three modes approach 2.3, 4.55 and 5 GHz, respectively. With this information, it is now worth comparing the A6 6-vane relativistic magnetron with the cavity magnetron examined in Sect. 3. It is worth noting that both Palevsky et al. [83] and Chen et al. [84] attributed these their largely to the radial and azimuthal components of the electric field for the π-mode, 2π-mode and $5\pi/3$-mode near the cathode are comparable in magnitude so leading to strong coupling of these modes to the circulating charge-space.

Table 3 details the diode aspect ratio and mode selection for each cavity magnetron. Column 4 in Table 3 lists the references where the physical cavity details and identified modes can be found. Note the Leningrad magnetron results are taken from particle-in-cell simulations by Andreev and Hendricks [85]. The rising-sun magnetron is not included in this comparison as the dual cavity aspect ratio (r) determines mode selection and competition see Sect. 5.3. The tabulated data reveals that the Leningrad, Birmingham and the two GEC magnetrons have σ values of 2 or less and only exhibit the π-mode under normal operation. It is only when the anode-cathode gap becomes small with $\sigma > 2$ mode competition is present.

Phase locking, rather than frequency locking, of two or more relativistic magnetrons to extended power level between 1 and 3 GW was demonstrated by Benford et al. in 1989 [6]. Here the two magnetrons were connected together using short length of waveguide that has a variable plunger for phase matching. This work is of direct interest to researchers in the field of chaos as the author's state: 'there is a limitation on the connector length to avoid chaos'.

6 Frequency Stability and Noise

So far the paper has focused on the magnetron's mode of operation which has been shown to depend upon the space-charge spoke interaction with the periodic structure of the anode. In this section the magnetrons ageing process, frequency pulling and cathode noise is now considered. To consider these processes the single charge space-charge assumption is modified to one of low density plasma where ions are formed from gas ions impurities which was first reported in experimental and theoretical detail by Langmuir in 1913 [22]. To understand spectral instabilities externally coupling at the electrical input and output of the magnetron oscillator needs also to be considered.

6.1 Cavity-Magnetron Aging

Like other electronic valves the performance of the magnetron can suffer due to gas impurity related instabilities which can cause arcing between the magnetron cathode and anode [11]. The resulting build-up of contaminant deposit is illustrated Fig. 10 and [86].

Arcing can have a number of origins, including: (1) incomplete evacuation of the valve during the manufacturing stage (2) virtual leaks around the electrodes, and (3) outgassing of material used in the construction of the valve. To prevent this malfunction the process of magnetron aging is used to 'clean up' the valve at the end of the manufacturing stage or performed as a prescheduled preventive maintenance procedure to prolong the life of the valve in service. The aging process takes the form of a 'bake-out' procedure using a sequence of increasing higher anode voltage steps, where each step initiates arcing and is maintain until the arcing ceases. This process is repeated until magnetron reaches its full specified operating voltage without arcing [68].

Fig. 10 Cut-away of a process aged (7000 h) Hitachi magnetron showing build-up of black deposit on the internal surface of the probe-antenna [86]

6.2 *Frequency Pulling*

Simply put, frequency pulling occurs when an external electromagnetic force acts upon the magnetron probe-antenna that is built into a waveguide structure and where the force, once passed the probe-antenna, alters the resonator- interaction coupling. Hence the magnetrons output frequency and power vary concurrently [87–89]. Normally the pulling figure is defined in terms of voltage standing wave ratio (VSWR) and is measured as the maximum change in output frequency that results when an external, fixed amplitude mismatch, located in the output waveguide, is moved through a distance of one half wavelengths relative to the magnetron. [*N.B. the magnetron pushing is a very different effect and is due to modulation of the magnetron power supply*].

6.3 *Cavity-Magnetron Cathode Noise*

In order to address cathode noise the work of the Welch and Dow [90] and Stewart [91] is considered. They postulated that each spoke of the electromagnetic slow-wave structure is composed of electrons randomly emitted from the heated oxide-coated cathode. Under these conditions the velocity of the electrons will be random and consequently be subjected to random fluctuations about a mean position. In the time domain these fluctuations are measured as a spoke jitter (or in the frequency domain measured as phase noise). This jitter or phase noise may be considered to be made up of both thermal noise as well as noise originating from the ionization of atoms of the cathode oxide coating by electron back-bombardment. Both of these processes may be amplified by the continuous interchange of energy between the dc field of the space-charge region and the electromagnetic field that is present within the interaction space. Thus spoke jitter (phase noise) leads to modulation of the magnetron's oscillator frequency.

Since the work of Welch and Stewart [90, 91] it has been found that the jitter (phase noise) bandwidth that can be significantly reduced by a number of means: three of which are presented here. Firstly, the spectral noise is a reduced using a cavity magnetron which is operated by a dc stabilized power supply and whose filament current is turned-off when the oscillations start [92]. Azimuthally varying axial magnetic fields may also be used to reduce spectral noise at all anode currents, but is particularly significant at low current near the start-oscillation condition [93]. Thirdly, metal shielding of the HV input side of the cathode can also reduce spectral noise [94].

7 Summary

This work has reviewed the magnetron electrical engineering and physics development; from the Barkhausen-Kurz oscillator to the relativistic cavity- magnetron using both US patent publications and peer-reviewed papers. As regards to the US patent office publications it is found to be a rich source of scientific and commercial detail. It is argued that the Barkheausenn-Kurz oscillator and the Hall magnetron provided little advantage over the triode valve in terms of increased frequency and output power. Between the 1920s and 1940s the magnetron design changed from one of open composite framed electrodes to one of solid enclosure resonator cavities circumferentially spaced around a cylindrical interaction space with an electron emitting cathode at its centre, thus allowing a wide range of frequencies to be supported within the coupled structure. However the development had very different design solutions depending upon the country of origin where the work was performed. The Sino-Japan, war (1931–1945), the war in the pacific (1941–1945) and in Europe (1939–1945) all had the specific influences that ultimately shaped today's magnetron designs (rising-sun cavity-magnetron, coaxial cavity-magnetron, relativistic cavity-magnetron and packaged cavity-magnetron).

It is concluded from this review that while military applications shaped magnetron development in the early part of the 20th century, as the technology developed the applications became increasingly focused on civilian use. Applications included medical therapeutic purposes (diathermy), commercial food preparation, and microwave heating as well as for its use in the plasma treatment of materials.

From a lumped-element EEM view point, and given the correct $\boldsymbol{E}\times\boldsymbol{B}$ conditions, both the frequency and output power of the cavity-magnetron is determined by the physical dimension of the resonator enclosures together with the reactive effect of any perturbations to the inductive or capacitive portion of the EEM. The coaxial cavity-magnetron differs in that the torroidal outer coaxial cavity controls the electromagnetic coupling into the slow-wave structure of the anode-cathode cavity. Excluding the coaxial cavity-magnetron, three π-mode controlling techniques have been described: strapping, alternating cavity structure within the rising-sun cavity-magnetron and diode aspect ratio in the relativistic cavity-magnetron. When each of these three techniques are properly applied the π-mode is automatically synchronises to the space-charge spoke pattern within interaction space, so removing mode competition.

A further aim of this work is to identify when and where mode competition is present within the magnetron family. This review has highlighted four (but not limited to) possible candidatures. The first and second candidates relate to the cavity-magnetron and are: the start-up period where mode competition, due to the rapidly growing space-charge, is most evident; the second is the ageing process that is characterised by internal surface erosion-carbonisation and gas impurity release. Candidate three is found to the sun-rise cavity-magnetron. In this case the N/2 – 1 competes with the stable π-mode the due resonator ratio (r_1) is <1.6 and, while for r_1 values >2 mode competition also occurs. The fourth candidate of interest is found in the A6 6-vane relativistic cavity-magnetron which can operate in the π-mode, 2π-mode and 5π/3-mode when the diode aspect ratio (σ) is > 2.

It has been shown that mode mathematical models have been developed for the cavity-magnetron family in the 1940–2010s The models include the Hartree component, or harmonic, and magnetron aging process. It is therefore reasonable to postulate that mode competition models may have a role in chaos theory when applied to natural and man-made coupled resonator networks.

Acknowledgements The authors would like to acknowledge support of the SFI funded I-Form Advanced Manufacturing Research Centre.

References

1. N.F. Alekseev, D.D. Malairov, Generation of high-power oscillations with a magnetron in the centimeter band. J. Tech. Phys. (USSR) **10**, 1297–1300 (1940)
2. J.T. Randell, H.A. Boot, High-frequency electrical oscillator. US 2,542,966 (issued Feb 20, 1951)
3. W.W. Hansen, R.D. Richtmeyer, On resonators suitable for klystron oscillators. J. Appl. Phys. **10**, 189–199 (1939)
4. W.W. Hansen, A type of electrical resonator. J. Appl. Phys. **9**(10), 654–663 (1938)
5. R.J. Collier, J. Feinstein, Magnetrons. US 2,854,603 (issued Sept 30, 1958)
6. J. Benford, H. Sze, W. Woo, R.R. Smith, B. Harteneck, Phase locking of relativistic magnetrons. Phys. Rev. Lett. **62**(8), 969–971 (1989)
7. Y. Shiyu, C. Jingzhong, T. Jianhui, L. Yaoting, A kind of magnetron cavity used in rubidium atomic frequency standards. J. Semicond. **32**(12), 122001 (2011)
8. C. Stefanucci, T. Bandi, F. Merli, M. Pellaton, C. Affolderbach, G. Mileti, A.K. Skrivervik, Compact microwave cavity for high performance rubidium frequency standards. Rev. Sci. Instrum. **83**, 104706 (2012)

9. R.G. Andrzejak, G. Ruzzene, I. Malvestio, Generalized synchronization between chimera states. Chaos **27**, 053114 (2017)
10. D. Dudkowski, Y. Maistrenko, T. Kapitaniak, Different types of chimera states: an interplay between spatial and dynamical chaos. Phys. Rev. E **90**, 032920 (2014)
11. V.J. Law, D.P. Dowling, Electronic valve instabilities and mode. CMSIM J. **1**, 3–35 (2017)
12. R.G. Aandrezak. C. Rummel, F. Mormann, K. Schindler, All together now: analogies between chimera state collapses and epileptic seizures. Sci. Rep. **6**, 23000 (2016)
13. M. Leconte, Statistical study of magnetron patents in the early years of electronics between 1920 and 1945; heuristic focusing around the discovery of the cavity magnetron, in *Proceedings of CAVMAG 2010* (IEEE, 2010), pp. 11–16. https://doi.org/10.1109/cavmag.2010.5565570
14. R.M. Page, *The Origin of Radar* (Anchor Books, Doubleday Garden City, New York, 1962), p. 66
15. *Proceedings of the First International Workshop on Cross-field Devices*, Michigan, USA, 15–16 Aug 1995
16. *Proceedings of CAVMAG 2010, International Conference on the Origins and Evolution of the Cavity Magnetron*, Bournemouth, UK, 19–20 Apr 2010
17. L. de Forest, Device for amplifying feeble electrical currents. U.S. 841,387 (issued Jan 15, 1907)
18. J.A. Fleming, On the conversion of electric oscillations into continuous currents by means of a vacuum valve. Proc. R. Soc. Lond. **74**, 476–487 (1904–1905)
19. J.A. Fleming, Instrument for converting alternating electric currents into continuous currents. US. 803,684 (issued Nov 7, 1905)
20. E.W. Armstrong, Some recent development in the audion receiver. Proc. IEEE **3**(3), 215–247 (1915)
21. E.W. Armstrong, Some recent development in the audion receiver. Reprinted in Proc. IEEE. **85**(4), 658–697 (1997)
22. E.V. Appleton, M.A.R. Barnett, On some direct evidence for downward atmospheric reflection of electric rays. Proc. R. Soc. Lond. A **109**(752), 621–641 (1925)
23. I. Langmuir, The effect of space charge and residual gases on thermionic currents in high vacuum. Phys. Rev. **2**(6), 450–486 (1913)
24. H.G. Barkhaushe, K. Kurz, Die kurzesten, in Vakuumrohren herstellbaren Wellen. Phys. Ztg. **21**, 1–6 (1920)
25. A.W. Hull, The magnetron. JAIEE **40**(9), 715–723 (1921)
26. N.A. Žáček, New method of generating undamped oscillations. J. Cultivation Math. Phys. **53**, 378–380 (1924)
27. E. Habann, Eine neue generatorröhr. Z. Hochfrequenztechnik **24**, 115–120 and 135–141 (1924)
28. K. Okabe, Production of intense extra-short radio waves by a split-anode magnetron. JIEEJ **48**(474), 284–290 (1928)
29. K. Posthumus, Oscillations in a split anode magnetron. Wirel. Eng. **12**, 126–132 (1935)
30. V.J. Law, Plasma harmonic and overtone coupling, in *Handbook of Applications of Chaos theory*, Chap. 20, ed. by C.H. Skiadas, C. Skiadas (Chapman and Hall/CRC Press, Taylor and Frances, 2016), pp. 405–422. ISBN-10: 1466590432
31. E.W. Gill, K.G. Britton, The action of a split-anode magnetron. JIEE **11**(32), 127–134 (1936)
32. Discussion on the action of a split-anode magnetron. JIEE **9**(476), 224–226 (1936)
33. G. Heller, The magnetron as a generator of ultra short waves. Philips Tech. Rev. **4**(7), 89–197 (1939)
34. H.E. Hollmann, Magnetron. US. 2,123,728 (issued July 12, 1938)
35. A.L. Samuel, Electron discharge device. US. 2,063,342 (issued Dec 8, 1936)
36. E.C.S. Megaw, The high-power pulsed magnetron: a review of early developments. JIEE Part IIIA Radiolocation **93**(5), 977–984 (1946)
37. J. Barbeen, W.H. Brattain, The transistor, a semi-conductor triode. Phys. Rev. **74**, 230–231 (1948)
38. K. Nosich, I.A. Tishchenko, Development of the first soviet three-coordinate L-band pulsed radar in kharkov before WWII. IEEE Antennas Propag. Mag. **4**(3), 29–48 (2001)

39. M. Elie, H. Gutton, J.H. Jean, M. Ponte, Détection d'obstacles à la navigation sans visibilité [Detection of obstacles in blind navigation]. Bulletin de la Société Française des Electriciens. 5ème série tome **IX**(100), 345–353 (1939)
40. M. Elie, H. Gutton, J.H. Jean, M. Ponte, System for object detection and distance measurement. US 2433838 (issued Jan 6, 1948)
41. K. Shimoda, Theory of split-anode magnetrons by Sin-itiro Tomonaga. AAPPS Bull. **16**(2), 17–22 (2006)
42. S. Nakajima, Japanese radar development prior to 1945. IEEE Antennas Propag. Mag. **34**(6), 17–22 (1992)
43. R.E. Edwards, Historical perspective on magnetron development, in *Proceedings of the First International Workshop on Cross-Field Devices*, Michigan, USA (1995), pp. 41–50
44. K. Frizz, Magnetron. US. 2,111,263 (issued Mar 15, 1938)
45. H. Gutton, S. Berline, Magnetron oscillator and detector. US. 2,147,159 (issued Feb 14, 1939)
46. A.A. Siddiqi, The Rockets' Red Glare: technology, conflict, and terror in the soviet union. Technol. Cult. **4**(3), 470–501 (2003)
47. N.F. Alekseev, D.D. Malairov, I.B. Bensen, Generation of high-power oscillations with a magnetron in the centimeter band. Proc. IRE **32**, 136–139 (1944)
48. H.A.H. Boot, J.T. Randall, Historical notes on the cavity magnetron. IEEE Trans. Electron Devices **23**(7), 724–729 (1976)
49. R.H. Varian, S.F. Varian, A high frequency oscillator and amplifier. J. Appl. Phys. **10**, 321–327 (1939)
50. R.W. Burns, Communications: an international history of the formative years. Chap 22 (IEE, London, UK, 2004), pp, 592–593. ISBN 0863413307
51. W.W. Hansen. High efficiency resonant circuit. US 2,190,712 (issued Feb, 1940)
52. Y. Blanchard, G. Galati, P. van Genderen, The cavity magnetron: not just a british invention. IEEE Antennas Propag. Mag. **55**(5), 244–254 (2013)
53. A.E. Austin, Precursors to radar—the Watson-Watt Memorandum and the Daventry experiment. IJEEE **36**, 365–372 (1999)
54. W. Froncisz, J.S. Hyde, Microwave resonator. US 4,446,429 (issued May 1, 1984)
55. D.R. Hartree, Mode selection in a magnetron by a modified resonance criterion. CVD Report Mag. **17** (Manchester University)
56. V.J. Law, S.D. Anghel, Compact atmospheric pressure plasma self-resonant drive circuits. J. Phys. D Appl. Phys. **45**(7), 075202 (2012)
57. R.H. Levy, Two new results in cylindrical diocotron theory. Phys. Fluids **11**(4), 920–921 (1968)
58. K.S. Fine, Simple theory of a nonlinear diocotron mode. Phys. Fluids **B4**(12), 3981–3984 (1992)
59. S. Riyopoulos, Magnetron instability in the low-space-charge limit. Phys Rev. Letts. **81**(14), 3026–3029 (1998)
60. R.I. Wilkinson, Short survey of Japanese radar-I. Electr. Eng. **65**, 455–463 (1946)
61. R.I. Wilkinson, Short survey of Japanese radar-II. Electr. Eng. **65**, 370–377 (1946)
62. H. Döring, Microwave tube development in Germany from 1920–1945. Int. J. Electron. **70**(5), 955–978 (1991)
63. G. Galati, On the Italian contribution to radar, in *European Radar Conference (EuRAD)* (2014), pp. 37–40
64. P.L. Spencer, Electron discharge device. US. 2,466,060 (issued Apr 5, 1949)
65. J. Sayers, High-frequency electrical oscillator. US 2,546,870 (issued Mar 27, 1951)
66. E.C. Okress, R.R. Reed, Echelon strapping system. US. 2,785,340 (issued Mar 12, 1957)
67. W.J. Dodds, Electron discharge device of the cavity resonator type. US. 2,607,019 (issued Aug 12, 1952)
68. A.A. Flowers, System for automatically aging of magnetrons and suppression of arcing thereof. US. 2,804,365 (issued Aug 27, 1957)
69. P.L. Spencer, Method of treating foodstuffs. US. 2495429 (issued. Jan 24, 1950)
70. J.F. Corbani, Cathode sputtering apparatus. US. 3,878,085 (issued Apr 5, 1975)
71. K. Suzuki, et al., Plasma etch device. US. 4,101,411 (issued Jul 18, 1978)

72. A. Ribner, Microwave plasma etching machine and method of etching. US 4,804,431 (issued Feb 14, 1989)
73. V.J. Law, D. Tait, Microwave plasma cleaning of ion implant ceramic insulators. Vacuum **49**(4), 273–278 (1998)
74. S. Millman, A.T. Nordsieck, The rising sun magnetron. J. Appl. Phys. **19**, 156–165 (1948)
75. N.M. Kroll, W.E. Lamb Jr., The resonant modes of the rising sun and other unstrapped magnetron. J. Appl. Phys. **19**, 166–186 (1948)
76. A.V. Hollenberg, N. Kroll, S. Millman, Rising sun magnetrons with large numbers of cavities. J. Appl. Phys. **19**, 624–635 (1948)
77. R.R. Moats, Interaction of modes in magnetron oscillators. Ph.D. Thesis, MIT, 1950
78. B. Van de Pol, On oscillation hysteresis in a triode generator with two degrees of freedom. Philos. Mag. **4**, 700–719 (1922)
79. D.C. Buck, Ferrite tuned coaxial magnetron. US. 3,333,148 (issued Dec 12, 1967)
80. D.E. Blank, G. Thornber, P. Wyann, Tunable coaxial magnetron. US. 3,590,312 (issued Jun 29, 1971)
81. G. Bekefi, T.J. Orzechowaski, Giant microwave bursts emitted from a field-emission, relativistic-electron beam magnetron. Phys. Rev. Lett. **37**(6), 379–381 (1976)
82. G. Bekefi, T.J. Orzechowaski, Relativistic electron beam cross-field device. US. 4,200,821 (issued Apr 29, 1980)
83. A. Palevsky, G. Bekefi, Microwave emission from pulsed, relativistic e-beam diodes. II. The multiresonator magnetron. Phys. Fluids. **22**, 986–996 (1979)
84. H.-W. Chen, C. Chen, R.C. Davidson, Numerical study of relativistic magnetrons. J. Appl. Phys. **73**(11), 7053–7060 (1993)
85. A.D. Andreev, K.J. Hendricks, First multi-cavity magnetrons were built in NII-9, Leningrad during the spring of 1937 *PIC Simulations of the First 4-Cavity S-Band CW Magnetron* (2010). https://doi.org/10.1109/cavmag.2010.5565560
86. V.J. Law, Knowledge domain information and the impact on process control. IUVSTA #49 Workshop, Dublin City University, Ireland, 4–6 Sept 2006. CD-rom published by the organizing Committee
87. A. Pagliarani, A.J. Kenyon, N.F. Thornhill, E. Sirisena, K. Lee, V.J. Law, Process harmonic pulling in a RIE plasma-tool. Electron. Lett. **42**(2), 120–121 (2006)
88. V.J. Law, Process induced oscillator frequency pulling and phase noise within plasma systems. Vacuum **82**(6), 630–638 (2008)
89. V.J. Law, N. Macgearailt, Visualization of a dual frequency plasma etch process. Meas. Sci. Technol. **18**(3), 645–649 (2007)
90. H.W. Welch, W.G. Dow, Analysis of synchronous conditions in the cylindrical magnetron space charge. J. Appl. Phys. **22**(4), 433–438 (1951)
91 J.L. Stewart, Theory of frequency modulation noise in tubes employing phase focusing. J. App. Phys. **26**(4), 409–413 (1955)
92. T. Mitani, N. Shinohara, H. Matsumoto, K. Hashimoto, Improvement of spurious noises generated from magnetrons driven by DC power supply after turning off filament current. IEICE Trans. Electron. **E86-C**(8), 1556–1563 (2003)
93. V.B. Neculaes, R.M. Gilgenbach, Y.Y. Lau, M.C. Jones, W.M. White, Low-noise microwave oven magnetrons with fast start-oscillation by azimuthally varying axial magnetic fields. IEEE Trans. Plasma Sci. **32**(3), 1152–1159 (2004)
94. T. Mitani, N. Shinohara, H. Matsumoto, M. Aiga, N. Kuwahara, T. Ishii, Noise-reduction effects of oven magnetron with cathode shield on high-voltage input side. IEEE Trans. Electron Devices **53**(8), 1929–1935 (2006)

The Fokker-Planck Equation and the First Exit Time Problem. A Fractional Second Order Approximation

Christos H. Skiadas and Charilaos Skiadas

1 The Stochastic Model

Following an approach in stochastic analysis we assume that the state $S = S(t)$ of an individual part of a stochastic system at time t follows a stochastic process of the form:

$$dS(t) = h(t)dt + \sigma(t)dW(t), \tag{1}$$

where $h(t)$ is the drift coefficient or the infinitesimal mean and $\sigma(t)$ the variance parameter or the infinitesimal variance or the diffusion coefficient and $W(t)$ the standard Wiener process. The latter is a good alternative to reproduce a stochastic process of Brownian motion type that is a random process to account for the random changes of our system. Accordingly an equation of the last type can model the time course of a complex system as are several complicated machines or automata. The last equation is immediately integrable provided we have selected appropriate initial conditions as $S(t = 0) = S(0)$.

$$S(t) = S(0) + \int_0^t h(s)ds + \int_0^s \sigma(s)dW(s), \tag{2}$$

This equation form gives a large number of stochastic paths for the health state $S(t)$ of the individual parts of the system. However, it should be noted that these paths following a random process with drift are artificial realizations that can not calculated in the real life for a specific part of the system else we have a perfect

C. H. Skiadas (✉)
ManLab, Technical University of Crete, Chania, Crete, Greece
e-mail: skiadas@cmsim.net

C. Skiadas
Department of Mathematics and Computer Science, Hanover College, Hanover, Indiana, USA
e-mail: skiadas@hanover.edu

C. H. Skiadas (ed.), *Fractional Dynamics, Anomalous Transport and Plasma Science*, https://doi.org/10.1007/978-3-030-04483-1_3

inspection system estimating the state of the particles in real time; that is impossible so far. However, we can find methods to estimate special characteristics of this system of stochastic paths provided by the last equation as is the summation of infinitesimal mean that is the mean value $H(t)$ of the systems state over time given by

$$H(t) = S(0) + \int_0^t h(s)ds, \tag{3}$$

Even so it is not feasible to calculate $H(t)$ immediately. Fortunately there are several theoretical approaches to find the time development of $H(t)$ from the advances in physics, mathematics and applied mathematics, and probability and statistics. The first approach is by observing that as $H(t)$ expresses the State of a large ensemble of the system's elements, it should be a declining function of time or better age.

2 General Solution

To find the appropriate form of $H(t)$ a series of delicate mathematical calculations are needed. We preferred estimates leading to closed form solutions thus providing important and easy applied tools for scientists from several fields and those non familiar with stochastic theory methodology and practice. The important steps leading to the final forms for estimating $H(t)$ are given in the following.

The first step is the formulation of the transition probability density function $p(S, t)$, that is a function expressing the probability for the health state S of an individual to move from one point at time t to the next. This is achieved by calculating a Chapman-Kolmogorov equation for the discrete case. However, we use here the continuous alternative of this equation that is the following Fokker-Planck equation:

$$\frac{\partial p(S,t)}{\partial t} = -h(t)\frac{\partial [p(S,t)]}{\partial S} + \frac{1}{2}[\sigma(t)]^2 \frac{\partial^2 [p(S,t)]}{\partial S^2}, \tag{4}$$

The related information can be found in Adriaan Fokker doctoral dissertation in Leiden University in 1913 [1] and in his paper 1914 [2]. Fokker in his disertation presents the related equation in connection with Max Planck theory in a specific chapter. Latter on Max Plack presented the Fokker-Planck equation in his paper in 1917 [4].

2.1 *Fractional Forms of the Fokker-Planck Equation*

The main fractional forms of the Fokker-Planck equation are the following

$$\frac{\partial p(S,t)}{\partial t} = -h(t)\frac{\partial [p(S,t)]}{\partial S} + \frac{1}{2}[\sigma(t)]^2 \frac{\partial^\alpha [p(S,t)]}{\partial S^\alpha}, \tag{5}$$

where α is the fractional parameter

In several problems in physics and engineering it is preferred to set $D = \frac{1}{2}\sigma^2$ to obtain the form

$$\frac{\partial p(S,t)}{\partial t} = -h(t)\frac{\partial [p(S,t)]}{\partial S} + D\frac{\partial^a [p(S,t)]}{\partial S^a}, \tag{6}$$

The selection of the diffusion parameter D became famous from the Einstein paper in a case of diffusion problem with zero drift with a non-fractional (classical) equation of the form

$$\frac{\partial p(S,t)}{\partial t} = D\frac{\partial^2 [p(S,t)]}{\partial S^2}, \tag{7}$$

The other fractional form of the Fokker-Planck equation refers to a fractional derivative with respect to time

$$\frac{\partial^\gamma p(S,t)}{\partial^\gamma t} = -h(t)\frac{\partial [p(S,t)]}{\partial S} + \frac{1}{2}[\sigma(t)]^2\frac{\partial [p(S,t)]}{\partial S}, \tag{8}$$

where γ is the fractional parameter.

The literature for the solution of these fractional Fokker-Planck equations is already quite large (see Sun et al. [12]). A significant part of this literature is included in the chapters of this book along with important applications.

2.2 Solution of the Fokker-Planck Equation

The solution of the classical Fokker-Planck partial differential equation is highly influenced by the boundary conditions selected. Here this partial differential equation for S and t is solved for the following appropriate boundary conditions (see Janssen and Skiadas [3])

$$p(S(t), t_0; S_0, t_0) = \delta(S(t) - S_0), \tag{9}$$

$$\frac{\partial p[S(t), t_0; S_0, t]}{\partial S(t)} \to 0 \quad \text{as} \quad S(t) \to \pm\infty \tag{10}$$

For the solution we use the method of characteristic functions. The characteristic function $\phi(S,t)$ is introduced by the following equation

$$\phi(S,t) = \int_{-\infty}^{+\infty} p(S,t; S_0, t_0)\exp(isS)ds, \tag{11}$$

Integrating by parts and using the Fokker-Plank equation we arrive at

$$\frac{\partial \phi}{\partial t} = ish(t)\phi - \frac{1}{2}[\sigma(t)]^2 s^2 \phi, \tag{12}$$

which with the initial conditions proposed

$$\phi(s, t_0) = \exp(isS_0), \tag{13}$$

is solved providing the following expression for ϕ

$$\phi(s, t_0) = \exp\left[is\left[S_0 + \int_{t_0}^{t} h(t')dt'\right] - \frac{1}{2}s^2 \int_{t_0}^{t} \left[\sigma(t')\right]^2 dt'\right], \tag{14}$$

This is the characteristic function of a Gaussian with mean

$$\left[S_0 + \int_{t_0}^{t} h(t')dt'\right], \tag{15}$$

and variance

$$\left[\sigma(t')\right]^2 dt', \tag{16}$$

Considering Eq. 3 and $t_0 = 0$ the solution is

$$p(t) = \frac{1}{[2\pi \int_0^t [\sigma(s)]^2 ds]^{1/2}} \exp\left[-\frac{[H(t)]^2}{2\int_0^t [\sigma(s)]^2 ds}\right], \tag{17}$$

3 Specific Solution

When $\sigma(t) = \sigma$ a simple presentation of the transition probability density function during time is given by:

$$p(t) = \frac{1}{\sigma\sqrt{2\pi t}} \exp\left[-\frac{[H(t)]^2}{2\sigma^2 t}\right], \tag{18}$$

Having estimated the transition probability density function for the continuous process we can find the first exit time probability density function for the process reaching a barrier.

4 A First Approximation Form

Just after the introduction of the Fokker-Planck equation in 1913 it was possible to estimate the first exit or hitting time of a stochastic process from a barrier. Schrödinger [5] and Smoluchowsky [11] solved the problem publishing two independent papers in the same journal issue in 1915. The drift should be linear of the form

$$H(t) = l - bt, \tag{19}$$

where l and b are parameters. Then the resulting distribution function $g(t)$ is known as the Inverse Gaussian of the form

$$g(t) = \frac{l}{t}p(t) = \frac{l}{t}\frac{1}{\sigma\sqrt{2\pi t}}\exp\left[-\frac{[l-bt)]^2}{2\sigma^2 t}\right], \tag{20}$$

Figure 1 (left) illustrates the Inverse Gaussian model. The linear drift is presented by a red heavy line whereas the stochastic paths appear as light lines. The confidence intervals are also presented by dashed lines. The inverse Gaussian Distribution is illustrated in Fig. 1 (right) as well as a fit curve to data (Data from the Carey Medflies).

The Inverse Gaussian as expressed in the last equation is a convenient form to find a generalization for a smooth nonlinear drift $H(t)$. From Fig. 2 we consider a tangent approximation in the point $M(H, t)$ of the drift curve $H(t)$.

$$l(t) = H(t) - tH'(t), \tag{21}$$

Note that the minus sign $(-)$ accounts for the negative slope of the derivative $H'(t)$. Now the only needed is to replace l by $l(t)$ and $l - bt$ by $H(t)$ in the formula of the Inverse Gaussian to obtain to tangent or first approximation for the first exit time density of a general smooth drift.

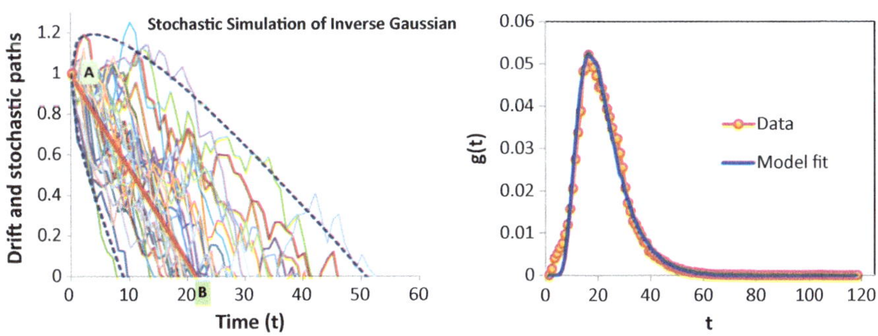

Fig. 1 (Left) Presentation of the inverse Gaussian with stochastic paths and (right) the inverse Gaussian distribution. Data from Carey Medflies

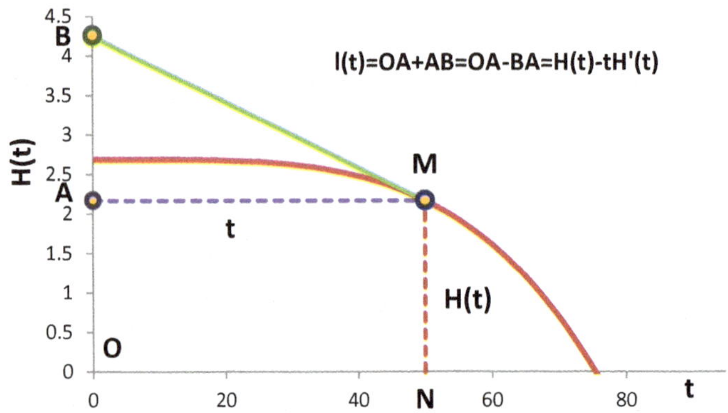

Fig. 2 Generalization of the inverse Gaussian

$$g(t) = \frac{|H - tH'|}{t} p(t), \tag{22}$$

By using the estimated $p(t)$ we arrive at the following form for the first exit time probability density function

$$g(t) = \frac{|H - tH'|}{\sigma\sqrt{2\pi t^3}} \exp\left[-\frac{[H(t)]^2}{2\sigma^2 t}\right], \tag{23}$$

The last formula is coming from a first approximation of the first exit time densities with good results in relatively simpler cases (see Skiadas and Skiadas [6–8]).

5 A Second Order Fractional Correction

Clearly the first order approximation error is smaller as the the drift $H(t)$ approaches linearity. For the other cases a second order approximation is needed by means of taking into account the second order derivative or even higher order derivatives. However, the second order derivative approach could be a good approximation provided that a fraction of this derivative is selected to account of the smaller or larger curvature of the drift $H(t)$. The resulting formula is the following

$$g(t) = \frac{1}{\sigma\sqrt{2\pi t}} \left[\frac{|H - tH'|}{t} + k\frac{t^2}{2} \frac{H''}{|H - tH'|}\right] \exp\left[-\frac{[H(t)]^2}{2\sigma^2 t}\right], \tag{24}$$

where the parameter k expresses the fraction of the second derivative needed. We take the quadratic term of a Taylor series expansion for $H(t)$ that is

$$H(t) = H(0) + tH' + \frac{t^2}{2}H'' + \cdots, \tag{25}$$

Evenmore the first order approximation $|H - tH'|$ is used as a normalising factor for the quadratic term.

5.1 An Interesting Application

We can arrive in a very interesting formula by selecting the following form for $H(t)$:

$$H(t) = l - (bt)^c, \tag{26}$$

where l, b, c are parameters

In our applications $H(t)$ expresses the health state of a population and thus the first exit time distribution $g(t)$ refers to the age of the population by means of $0 \leq t$. Accordingly the $g(t)$ is a half distribution and the resulting first and second order approximations are:

$$g(t) = \frac{2|l + (c-1)(bt)^c|}{\sigma\sqrt{2\pi t^3}} \exp\left[-\frac{[l - (bt)^c]^2}{2\sigma^2 t}\right], \tag{27}$$

$$g(t) = \frac{2}{\sigma\sqrt{2\pi t}}\left[\frac{|l + (c-1)(bt)^c|}{t} + \frac{kt^2 c(c-1)b^c t^{(c-2)}}{2|l + (c-1)(bt)^c|}\right] \exp\left[-\frac{[l - (bt)^c]^2}{2\sigma^2 t}\right], \tag{28}$$

As for fitting this formula and the previous simpler forms to data sets it is not possible to estimate the parameters of the model along with σ, two options are selected; that is to set $\sigma = 1$ and estimate l, b, c or to set $l = 1$ and estimate b, c, σ. The latter is very important when stochastic simulations are needed. It is also useful for applications on health state estimates. It was selected from Weitz and Fraser [13] for the application in Medflies and from Skiadas and Skiadas [6–10] for many applications.

The applications are done in actual data (mortality) sets or on a logarithmic transformation of the data providing better information for the first period of the human lifespan. The first order approximation (see Fig. 3) fail to cover the first period of the lifespan where strongly nonlinear forms appear. Instead, the fractional approximations with the second derivative (see Fig. 4) provide very good fitting to the data sets.

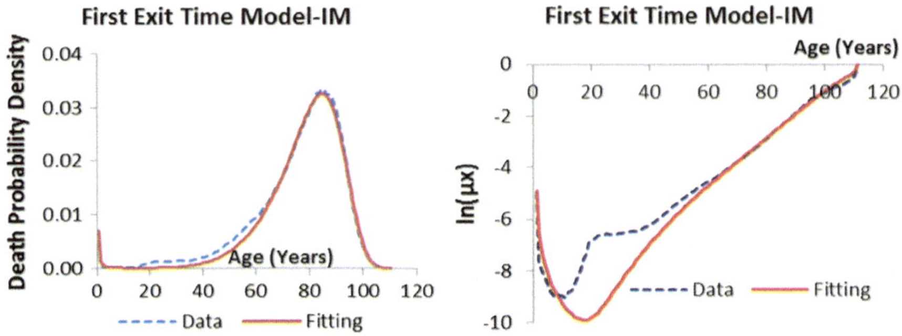

Fig. 3 (Left) First exit time distribution (data and fit curves for USA mortality data, male 2010) and (right) respective application with a logarithmic form of the data. First order approximation

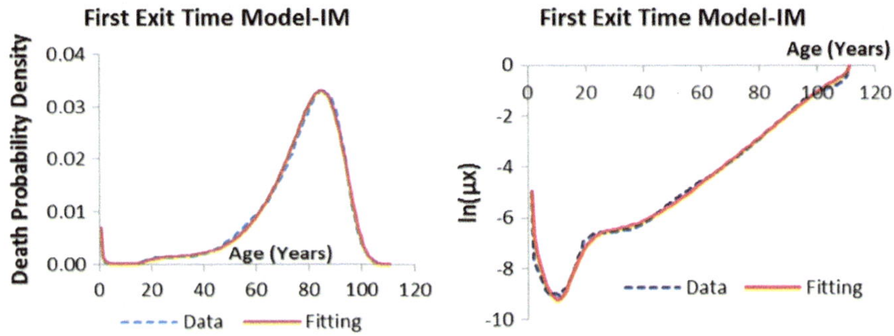

Fig. 4 (Left) First exit time distribution (data and fit curves for USA mortality data, male 2010) and (right) respective application with a logarithmic form of the data. Second order approximation

6 Summary and Conclusions

We have present a first exit time theory of a stochastic process. The general model is analytically derived according to the first exit time or hitting time theory for a stochastic process crossing a barrier. The derivation lines follow the transition probability densities from the Fokker-Planck equation. Then we find the probability density form and the first and second approximation of the first exit time densities. For the first approximation we obtain a generalization of the Inverse Gaussian whereas for the second approximation we apply a fractional approach to the second derivative by introducing a parameter k. We thus introduce another approach to apply a fractional theory. Instead to apply the fractional derivative theory to the Fokker-Plank equation we have solved a classical Fokker-Plank equation and then we have selected a fractional approach when introducing the second order derivative.

References

1. A. Fokker, *Over Brown'sche bewegingen in het stralingsveld, en waarschijnlijkheids-beschouwingen in de stralingstheorie*. Ph.D. Dissertation, Leiden University, The Netherlands, 1913
2. A. Fokker, Die mittlere Energie rotierender elektrischer Dipole im Strahlungsfeld. Annalen der Physik **348**(4. Folge 43), 810–820 (1914)
3. J. Janssen, C.H. Skiadas, Dynamic modelling of life-table data. Appl. Stoch. Models Data Anal **11**(1), 35–49 (1995)
4. M. Planck, *Über einen Satz der statistischen Dynamik und seine Erweiterung in der Quanten-theorie*, vol. 24 (Sitzungsberichte der Preussischen Akademie der Wissenschaften zu Berlin, 1917), pp. 324–341
5. E. Schrödinger, Zur theorie der fall- und steigversuche an teilchenn mit Brownsche bewegung. Physikalische Zeitschrift **16**, 289–295 (1915)
6. C. Skiadas, C.H. Skiadas, Development, simulation and application of first exit time densities to life table data. Commun. Stat.-Theory Methods **39**, 444–451 (2010)
7. C.H. Skiadas, C. Skiadas, The first exit time theory applied to life table data: the health state function of a population and other characteristics. Commun. Stat.-Theory Methods **43**, 1985–1600 (2014)
8. C.H. Skiadas, C. Skiadas, Exploring the state of a stochastic system via stochastic simulations: an interesting inversion problem and the health state function. Methodol. Comput. Appl. Probab. **17**, 973–982 (2015)
9. C.H. Skiadas, C. Skiadas, *Exploring the Health State of a Population by Dynamic Modeling Methods* (Springer, Chum, Switzerland, 2017)
10. C.H. Skiadas, C. Skiadas, *Demography and Health Issues: Population Aging* (Mortality and Data Analysis, Springer, Chum, Switzerland, 2018)
11. M.V. Smoluchowsky, Notiz ber die Berechnung der Browschen Molekular bewegung bei der Ehrenhaft-Millikanschen Versuchsanordnung. Physikalische Zeitschrift **16**, 318–321 (1915)
12. H.G. Sun, W. Chen, K.Y. Sze, A semi-discrete finite element method for a class of time-fractional diffusion equations. Phil. Trans. R. Soc. A **371**, 20120268 (2013)
13. J.S. Weitz, H.B. Fraser, Explaining mortality rate plateaus. Proc. Natl. Acad. Sci. USA **98**(26), 15383 15386 (2001)

Anomalous Diffusion by the Fractional Fokker-Planck Equation and Lévy Stable Processes

Johan Anderson and Sara Moradi

1 Introduction

Finding a proper kinetic description of dynamical systems with chaotic behaviour is one of the main problems in classical physics. Over the past two decades it has become obvious that much more complex behaviour than standard diffusion can occur in dynamical Hamiltonian chaotic systems. In principle, the orbits in dynamical systems are always theoretically predictable since they arise as solutions to simple system of equations such as Newton's equations however these orbits are sensitive to initial conditions. From the macroscopic point of view, the rapid mixing of orbits has been used as a motivation for assumptions of randomness of the motion, and the random walk models.

During the last century, classical random walk or Brownian motion was studied extensively and used as a paradigm to understand and describing the diffusion phenomenon. In a Brownian motion the mean value vanishes whereas the second moment or variance grows linearly with time $\langle \delta x^2 \rangle = 2Dt$. However, many phenomena exhibit anomalous diffusion where variance grows non-linearly in time such that $\langle \delta x^2 \rangle - 2Dt^{\alpha}$. There is no mechanism that inherently constrains $\lim_{\delta x, \delta t \to 0} \frac{\delta x^2}{\delta t}$ to be finite or non-zero. In more general terms, there are two limits of interest where the first is super-diffusion $\alpha > 1$ and the second is sub-diffusion with $\alpha < 1$. Such strange kinetics [1–6] may be generated by accelerated or sticky motions along the trajectory of the random walk [7]. Super-diffusivity can occur as a result of variation in the step length of the motion which gives rise to long-range correlations in the dynamics

J. Anderson (✉)
Department of Space, Earth and Environment, Chalmers University of Technology, 412 96 Göteborg, Sweden
e-mail: anderson.johan@gmail.com

S. Moradi
Laboratory for Plasma Physics - LPP-ERM/KMS, Royal Military Academy, 1000 Brussels, Belgium
e-mail: sara.moradi@ukaea.uk

C. H. Skiadas (ed.), *Fractional Dynamics, Anomalous Transport and Plasma Science*, https://doi.org/10.1007/978-3-030-04483-1_4

generated by the presence of anomalously large particle displacements connecting otherwise physically disjoint domains. Sub-diffusive properties has been studied in many different contexts where transport is often inhibited by sticky motion. Among sub-diffusive phenomena are that of holes in amorphous semiconductors where a waiting time distribution with long tails was introduced [8] and the sub-diffusive processes within a single protein molecule described by generalized Langevin equation with fractional Gaussian noise [9]. Turbulence and related anomalous diffusion phenomena have been observed in a wide variety of complex systems such as high energy plasmas, semiconductors, glassy materials, nanopores, biological cells, and epidemic proliferation.

In these situations, thus, kinetic descriptions which arise as a consequence of averaging over the well-known Gaussian and Poissonian statistics (for diffusion in space and temporal measures, respectively) seem to fall short in describing the apparent randomness of dynamical chaotic systems [1]. This is due to the restrictive assumptions of locality in space and time, and lack of long-range correlations that is the basis of these descriptions.

In magnetised plasmas, it is commonly accepted that turbulence is the primary cause of anomalous (i.e. elevated compared to collisional) transport. It has been recognized that the nature of the anomalous transport processes is dominated by a significant ballistic or non-local component where a diffusive description is improper. Super-diffusive properties are often found with $\alpha > 1$ such as the thermal and particle flux in the gradient region of the magnetically confined plasmas or in the Scrape-Off Layer (SOL) where the transport is dominated by coherent structures [10–21] and inherently non-local effects [22–28]. Moreover, analysis of measurements with Langmuir probes at different fusion devices of density and potential fluctuations have provided evidence to support the idea that these fluctuations are distributed according to Lévy statistics. This was illustrated for example in [29], where PDFs of the turbulence induced fluxes at the edge of W7-AS stellarator were shown to exhibit power law characteristics in contrast to exponential decay expected from Gaussian statistics. Furthermore, the experimental evidence of the wave-number spectrum characterized by power laws over a wide range of wave-numbers can be directly linked to the values of Lévy index α of the PDFs of the underlying turbulent processes. Lévy statistics describing fractal processes (Lévy index α where $0 < \alpha < 2$) lie at the heart of complex processes such as anomalous diffusion. Lévy statistics can be generated by random processes that are scale-invariant. This means that a trajectory will possess many scales, but no one scale will be characteristic and dominate the process. Geometrically this implies the fractal property that a trajectory, viewed at different resolutions, will look self-similar.

Understanding anomalous transport in magnetically confined plasmas is an outstanding issue in controlled fusion research. A satisfactorily understanding of the non-local features as well as the non-Gaussian probability distribution functions (PDFs) found in experimental measurements of particle and heat fluxes is still lacking [15, 16]. Fractional kinetics has been put forward for building a more physically relevant kinetic description for such dynamics.

Fractional kinetics is a powerful framework in describing anomalous transport processes exibiting Lévy statistics. It is able to reproduce key aspects of anomalous transport including the non-Gaussian self-similar nature of the PDFs of particle displacement, and the anomalous scaling of the moments. Also, the integro-differential nature of the fractional derivative operators allows the description of spatiotemporal nonlocal transport processes. In particular, in fractional diffusion, the local Fourier-Fick's law is replaced by an integral operator in which the flux at a given point in space depends globally on the spatial distribution of the transported scalar, and on the time history of the transport process. Using fractional generalizations of the Liouville equation, kinetic descriptions have been developed previously [30–32]. It has been shown that the chaotic dynamics can be described by using the Fokker-Planck (Fp) equation with coordinate fractional derivatives as a possible tool for the description of anomalous diffusion [33]. Previous papers on plasma transport have used models including a fractional derivatives on phenomenological premises [13, 34].

The goal of this work is to give an overview and new insights into the recent developments in modelling of the anomalous transport of charged particles in magnetised plasmas. Different approaches to the problem from solving Monte Carlo simulations of charged particles in the presence of α-stable Lévy fluctuations in an external magnetic field and linear friction, and numerical solution of Fractional Fokker-Planck Equation (FFPE) were developed [18–20, 34–36].

2 Modelling of Anomalous Diffusion by the Langevin Equation

One kinetic description in plasma physics is given by the Vlasov equation (a generalization of the Boltzmann equation, where the forces include not only the external ones but also the mean electromagnetic fields due to the plasma itself) that requires a 6 dimensional phase space (3 for positions and 3 for the velocities). Plasma dynamics is often dominated by collective motion which using the Vlasov description is effectively impossible to directly compute due to the rapid Larmor gyrations. Instead, an alternative formulation reducing the dimensionality of the problem is used, which describes the motion of the guiding centres of these rapidly rotating plasma particles in a 5D phase space [21]. This formulation, called gyro-kinetic theory, is commonly used in simulating plasma turbulence. The gyro-kinetic formulation however strongly rests on the assumption that the motion of charged particles following the magnetic field lines is local and therefore averaging over the Larmor gyration possible. In practice, the applicability of gyro-averaged models may be limited due to non-local and intermittent properties of the plasma induced by the collective effects of electromagnetic forces between the charged particles.

It is thus pertinent to explore the non-locality or fractal properties of charged plasma dynamics. Here the statistics of charged particle motion in the presence of α-stable Lévy fluctuations in a external magnetic field and linear friction is discussed in the framework of numerical Monte Carlo simulations. The Lévy noise was introduced to model the effect of non-Gaussian, intermittent electrostatic fluctuations. The statistical properties of the velocity moments and energy for various values of the Lévy index α were investigated as well as the role of Lévy fluctuations on the statistics of the particles' Larmor radii in order to examine potential limitations of gyro-averaging. Previously in Ref. [37] numerical work were limited to 2 dimensions, this is now expanded further by utilizing 3-dimensional simulations in a cylindrical magnetic field, where the statistics of the spatial displacements and Larmor radius are studied. The work is limited by neglecting memory effects and the Lévy noise was taken as white or delta correlated in time. One other aspect is that these simulations can be used to corroborate analytical models of fractional dynamics, which we will discuss later in this work.

We consider the motion of charged particles in a 3-dimensional magnetic field in a cylindrical domain in the presence of linear friction modelling collisional Coulomb drag and a stochastic electric field according to the Langevin equations

$$\frac{d\mathbf{r}}{dt} = \mathbf{v}, \tag{1}$$

$$\frac{d\mathbf{v}}{dt} = \frac{q_s}{m_s}\mathbf{v} \times \mathbf{B} - \nu\mathbf{v} + \frac{q_s}{m_s}\mathcal{E}. \tag{2}$$

Here q_s and m_s are the charge and mass of the particle species s, ν is the friction parameter and $\mathcal{E}$ is a 3-dimensional, homogeneous, isotropic turbulent electric field modelled as an stationary, uncorrelated stochastic process without memory following an α-stable distribution, $f(\alpha, \beta, \sigma, \eta)$, with characteristic exponent $0 < \alpha \leq 2$, skewness $\beta = 0$, variance $\sigma = 1/\sqrt{2}$, and mean $\eta = 0$. Here, we use the definition of $f(\alpha, \beta, \sigma, \eta)$ as described in Refs. [38–40].

A periodic straight cylindrical domain with period $L = 2\pi R_0$ is considered, with R_0 being the major radius, and we use cylindrical coordinates (r, θ, z). The magnetic field is a helical field of the form,

$$\mathbf{B}(r) = B_\theta(r)\,\hat{\mathbf{e}}_\theta + B_z\hat{\mathbf{e}}_z. \tag{3}$$

A constant magnetic field in z-direction, $B_z = B_0$, is assumed. The shear of the helical magnetic field, i.e. the dependence of the azimuthal rotation of the field as function of the radius, is determined by the q-profile, $q(r) = rB_z/(R_0 B_\theta)$, where

$$B_\theta(r) = \frac{B(r/\lambda)}{1 + (r/\lambda)^2}, \tag{4}$$

for which the q profile is

$$q(r) = q_0\left(1 + \frac{r^2}{\lambda^2}\right). \tag{5}$$

In terms of the flux variable,

$$\psi = \frac{r^2}{2R_0^2}, \tag{6}$$

q is a linear function of ψ. The numerical integration of Eqs. (1) and (2) is performed using a Runge-Kutta 4th order scheme (RK4) over the interval $[0, T]$. The time step for the RK4 integration is defined by partitioning the interval $[0, T]$ into N subintervals of width $\delta = T/N > 0$,

$$0 = \tau_0 < \tau_1 < \cdots < \tau_i < \tau_N = T, \tag{7}$$

with the initial conditions $\mathbf{r}_0$, and $\mathbf{v}_0$. We compute $\mathbf{r}_i$ and $\mathbf{v}_i$ for the subintervals with the time step of $dt = \delta/n$, and at every δ, we include the cumulative integral of the stochastic process using

$$d\mathbf{r}_i = \mathbf{v}_i dt \tag{8}$$

$$d\mathbf{v}_i = \left[\frac{q_s}{m_s}\mathbf{v}_i \times \mathbf{B} - \nu \mathbf{v}_i\right] dt + \mathbf{W} \tag{9}$$

where

$$\mathbf{W} = \frac{q_s}{m_s}\chi \sum_{\delta} (dt)^{(1/\alpha)}\mathcal{E}. \tag{10}$$

Here, using spherical coordinates, random samples in the $\mathcal{E}_\rho$ radial direction are generated with the α-stable random generator developed in Refs. [38–40], and two uniformly distributed angles θ and ϕ between $[0, 2\pi]$ are used. In Cartesian coordinates the components of the electric field are $\mathcal{E}_x = \mathcal{E}_\rho \sin\theta \cos\phi$, $\mathcal{E}_y = \mathcal{E}_\rho \sin\theta \sin\phi$, and $\mathcal{E}_z = \mathcal{E}_\rho \cos\theta$. $N_p = 10^4$ particles are considered, and the simulation time is $T = 500/\tau_c$ where $\tau_c = 2\pi/\Omega_c$ and $\Omega_c = |q_s|B_0/m_s$ is the gyration frequency. We explore the dependence of the particle motion on the index α of the Lévy fluctuations and the parameter $\epsilon = \chi/\nu$ where χ is the amplitude of the fluctuations and ν is the damping coefficient. The convergence in probability of Lévy driven stochastic differential Eqs. 1 and 2 have been discussed in [41] where a criteria is established.

Figures 1 and 2 show samples of the computed particle orbits and their respective velocities for $\alpha = 2$ and 1.5. As seen in the Fig. 1 (left), for the Gaussian process, the orbits tightly follow the helical field lines and their motion homogeneously cover the 3-d velocity space as seen in Fig. 2 (right). However, in the case of Lévy process with $\alpha = 1.5$, the gyro orbits deviate significantly from the magnetic field lines, and outlier events are observed to result in a significant increase in the velocity space covered by the particle's motion, as shown in Fig. 2 (right).

In general it is noted that the PDFs change from an exponential decay to a power law decay when the stochastic process is changed from a Gaussian to a Lévy process. This is an indication of induced non-local effects stemming from large fluctuations. Most often turbulence in plasmas exhibit a very large range of spatio-temporal scales. This leads to a formidable computational challenge, it is customary to use reduced descriptions based on spatial and/or temporal averaging of degrees of freedom that

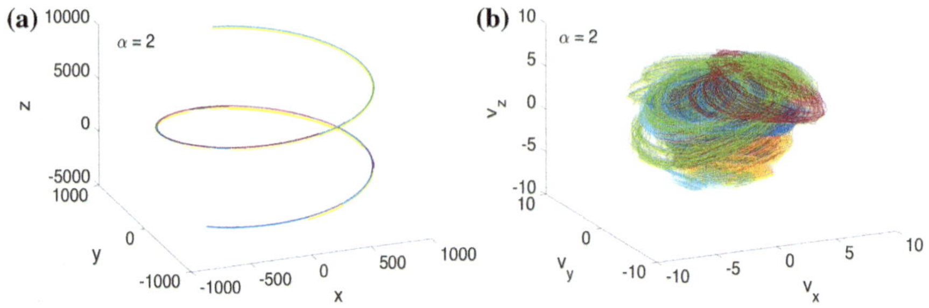

Fig. 1 (Left) Samples of computed particle orbits normalised to $\rho_L(0)$ and (right) their respective velocity for a Gaussian stochastic process, $\alpha = 2$, and $\epsilon = 100$

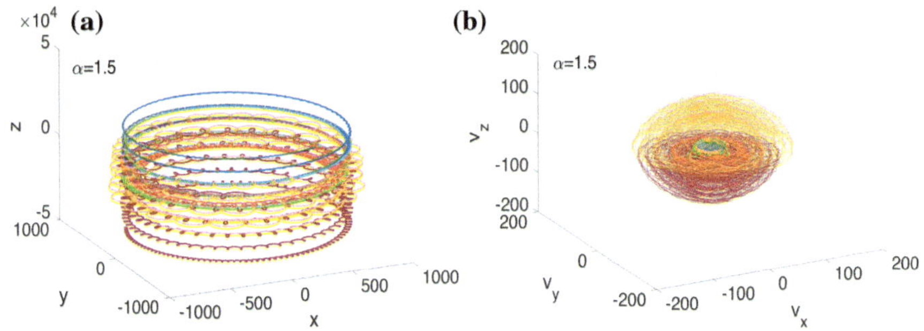

Fig. 2 (Left) Samples of computed particle orbits normalised to $\rho_L(0)$ and (right) their respective velocity for a Lévy stochastic process, with $\alpha = 1.5$, and $\epsilon = 100$

evolve on small spatial scales and/or fast time scales compared to the macroscopic scales of interest. One common example is the extensively used gyro-kinetic models, where it is assumed that $\rho_L/L \ll 1$ where ρ_L is the Larmor radius and L is the tokamak minor radius or a characteristic density gradient scale length. However, it is important to keep in mind that in a turbulent plasma where the fluctuations follow a Lévy process also the Larmor radius cannot be deterministcally determined it thus is a statistical quantity, $\langle \rho_L \rangle$, (where $\langle \cdot \rangle$ denotes ensemble average). On the other hand, in a plasma in Maxwellian equilibrium the PDF of Larmor radii is sharply peaked around the thermal Larmor radius. However, for plasmas where the PDF exhibits slowly decaying tails with a large amount of extreme events (i.e., particles with anomalously large Larmor radii) the situation is much less trivial. Especially, in the case of algebraic decaying PDFs, where statistical moments might not exists or converge it might not be possible to obtain a characteristic length scale of the process. The results obtained by these simulations indicate that when turbulent electrostatic fluctuations exhibit non-Gaussian Lévy statistics, gyro-averaging and guiding centre approximations may not be fully justified and full particle orbit effects should be taken into account. Investigations of scale free stochastic processes has attracted significant interest in plasma physics in particular but also in applied and basic sciences, see for example Refs. [42, 43] and references therein. For more details and in-depth discussions see Ref. [20].

3 Modelling Anomalous Transport

In the following we will discuss a fractional Fokker-Planck (FPP) approach to model anomalous processes, it thus pertinent to start with the theory of Brownian motion where an equation of motion for a colloidal particle in a background medium is in the form of a Langevin equation [44]

$$\frac{d\mathbf{v}}{dt} = -\nu\mathbf{v} + A(t) \tag{11}$$

Here, the Gaussian random noise, $A(t)$, represents the influence of the background medium by dynamical friction and a fluctuating part. The Gaussian white noise assumption is usually imposed in order to obtain a Maxwellian velocity distribution describing the equilibrium of the Brownian particle. This relation is discussed in Ref. [45] in relation to the connection between the Gaussian central limit theorem and classical Boltzmann-Gibbs statistics. However, the Gaussian central limit theorem is not unique and a generalization was done by Lévy [46], Khintchine [45] and Seshadri [47] using long tailed distributions.

The FFP can be derived in several ways where the most straightforward way is to consider long jumps, i.e., Lévy flights, which therefore allows for long tails in the equilibrium PDFs. We introduced the Lévy statistics into the Langevin equation thus yielding a FFP description. We follow the approach used by Fogedby [48, 49], Barkai [50] and Moradi [18, 19] where a FFPE with only a fractional velocity derivative in the presence of a constant external force is obtained as

$$\frac{\partial F}{\partial t} + \mathbf{v}\frac{\partial F}{\partial \mathbf{r}} + \frac{\mathbf{F}}{m}\frac{\partial F}{\partial \mathbf{v}} = \nu\frac{\partial}{\partial \mathbf{v}}(\mathbf{v}F) + D\frac{\partial^\alpha F}{\partial |\mathbf{v}|^\alpha}, \tag{12}$$

with $0 \leq \alpha \leq 2$. Here, the term $\frac{\partial^\alpha F}{\partial |\mathbf{v}|^\alpha}$ is the fractional Riesz derivative. The diffusion coefficient, D, is related to the damping term ν, according to a generalized Einstein relation [50]

$$D = \frac{2^{\alpha-1} T_\alpha \nu}{\Gamma(1+\alpha) m^{\alpha-1}}. \tag{13}$$

Here, T_α is a generalized temperature, and the force $\mathbf{F}$ represents the Lorentz force acting on the particles with mass m, and $\Gamma(1+\alpha)$ is the Euler gamma function.

Note that for Lévy type distribution functions, higher moments will diverge. Thus, it is of interest to define convergent statistical measures of the underlying process. We will employ the generalized q-moments or q-expectations as

$\langle v^p \rangle_q = \int dv F(v)^q v^p / \int dv F(v)^q$. The q-expectation can be a convergent moment of the distribution function although the regular moments diverges. This also gives us the opportunity to define a pseudo-energy that is always convergent.

In order to analytically investigate the effects of the fractional derivative on the diffusion we consider the force-less homogeneous one dimensional Fokker-Planck equation of the form,

$$\frac{\partial F}{\partial t} = \nu \frac{\partial}{\partial v}(vF) + D \frac{\partial^\alpha}{\partial |v|^\alpha} F. \tag{14}$$

The solution is found by Fourier transforming and treating the fractional derivative in the same manner as in Ref. [63] we find,

$$\frac{\partial \hat{F}}{\partial t} = -\nu k \frac{\partial}{\partial k} \hat{F} - D|k|^\alpha \hat{F}. \tag{15}$$

The stationary PDF is now readily obtained by integration and an inverse Fourier transform,

$$\hat{F}(k) = F_0 \exp\left(-\frac{D}{\nu\alpha}|k|^\alpha\right), \tag{16}$$

$$F(v) = \frac{F_0}{2\pi} \mathrm{Re}\left\{\int_{-\infty}^{\infty} dk \exp\left(-\frac{D}{\nu\alpha}|k|^\alpha + ikv\right)\right\}. \tag{17}$$

Due to the fractal form of the inverse Fourier transform analytical solutions of the PDF for the general case is difficult to obtain as closed functions, except in particular cases of $\alpha = 1.0$ and $\alpha = 2$ yielding a Lorentz distribution and a Gaussian distribution, respectively. Note, that it is possible to express the solutions in Fox's H-functions. Moreover it is interesting that Eq. (17) is equivalent to what was found in Ref. [32]. The PDF generated by Eq. (15) is,

$$F(v) = \frac{N}{(1 + \beta(q-1)v^2)^{1/(q-1)}}. \tag{18}$$

Here N is normalization factor. Furthermore, it is found that β is not representative of an inverse temperature of the system due to its non-equilibrium nature. In the rest of this paper we will study the solutions to the Eq. (17) in more detail.

In investigations of the anomalous character of transport a useful tool is the non-extensive statistical mechanics which provides distribution functions intermediate to that of Gaussians and Lévy distributions adjustable by a continuous real parameter q [51–53]. The parameter q describes the degree of non-extensivity in the system. Non-extensive statistical mechanics has a solid theoretical basis for analysing complex systems out of equilibrium where the total entropy is not equal to the sum of the entropies from each subsystem. However, it must be noted that Tsallis q-statistics is not unique. For systems comprised of independent or parts interacting through short-range forces the Boltzmann-Gibb statistical mechanics is sufficient however for systems exhibiting fractal structure or long range correlations this approach becomes unwarranted. Tsallis statistics is now widely applied e.g. to solar and space plasmas such as the heliosphere magnetic field and the solar wind [54–56]. The main topic of this paper is to evaluate the statistical properties in terms of Tsallis statistics dependent of the fractional index (α) in Eq. (17). We will start by numerically computing the PDFs with α as our free parameter. Subsequently, in order to statistically evaluate the numerically found PDFs in the fractal model we will determine the q-expectation and the Tsallis non-extensive entropy. Note that the regular statistical moments of the PDFs will not converge unless the PDFs are considered to have a finite compact support. Thus, we will now focus on solving Eq. (17) numerically, by computing the inverse Fourier transform and compare the found PDFs to previously derived analytical solutions Eq. (18).

In Fig. 3, the numerically found PDFs are shown (log-linearly) for $\alpha = 1.00$ (magenta line), $\alpha = 1.25$ (green line), $\alpha = 1.50$ (red line), $\alpha = 1.75$ (blue line) and $\alpha = 2.00$ (black line). Here, in this figure the diffusion coefficient over the dissipation is kept constant $D/\nu = 1.0$.

We note as the parameter α decreases, the normalized fourth moment (Kurtosis $= m_4/m_2^2 =$ the ratio of the fourth moment divided by the square of the standard deviation) of the symmetric PDF increases rapidly where PDFs become more and more peaked with elevated tails. Note that the distribution varies smoothly as α - parameter is decreased from a Gaussian distribution with $\alpha = 2.0$ down to the Lorentz distribution with $\alpha = 1.0$.

It has been shown, in Ref. [57], that using generalized statistical mechanics yielded PDFs that are of Cauchy-Lorentz form,

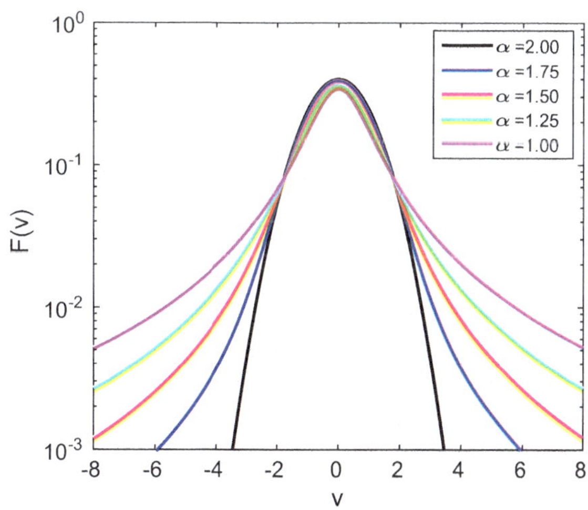

Fig. 3 The F(v) as a function of the velocity v for $\alpha = 2.00$ (black line), $\alpha = 1.75$ (blue line), $\alpha = 1.50$ (red line), $\alpha = 1.25$ (green line) and $\alpha = 1.00$ (magenta line)

$$F(v) = \frac{a}{(1 + b(q-1)v^2)^{1/(q-1)}}. \tag{19}$$

We note that this type of PDF exhibit power law tails that are significantly elevated compared to Gaussian or exponential tails. It is worthy noting that the precise analytical relation between the fractality index α and the non-extensivity parameter q is not entirely clear due to the non-uniqueness of the q-statistics. One possibility is the formal relation between the fractality index α and the non-extensivity q proposed by [57] as:

$$\alpha = \frac{3-q}{q-1}. \tag{20}$$

It can easily be shown that $F(v) \propto v^{-(\alpha+1)}$ as $v \to \infty$ which is fulfilled by Eq. 20. On the other hand in the limit of small v the exponential factor can be approximated as follows $e^{ikv} = 1 + ikv - \frac{1}{2}k^2v^2 + \cdots$, keeping only the even powers in the integral due to the symmetry, we find:

$$F(v) \approx \frac{F_0}{\pi}\frac{1}{\alpha}\left[\left(\frac{D}{\alpha\nu}\right)^{-1/\alpha}\Gamma\left(\frac{1}{\alpha}\right) - \left(\frac{D}{\alpha\nu}\right)^{-3/\alpha}\Gamma(\frac{3}{\alpha})\frac{v^2}{2!} + \cdots\right]. \tag{21}$$

Here Γ is the gamma function. Note that the last expansion in Eq. 21 approximates a Gaussian distribution function for small v for Eq. (20).

To further investigate the suitability of Eq. (20) with $q = 9/5$ for $\alpha = 3/2$ where we find good agreement over several orders of magnitude between the proposed analytically derived PDFs based on Eq. (20) and the numerically computed PDFs for the values used. Note that similar agreement is found for all values of $0.25 < \alpha < 1.5$, for a more elaborate discussion please consult Ref. [36].

While we follow the definition that any diffusive process that diverges from the form $\langle x^2\rangle(t) \propto t$ is called anomalous, in most cases we will deal with super-diffusion where $\langle x^2\rangle(t)$ may be divergent. In order to find a useful statistical measure of the super-diffusive or fractal process we introduce,

$$\langle v^2\rangle_q = \frac{\int_{-\infty}^{\infty} dv(F(v))^q v^2}{\int_{-\infty}^{\infty} dv(F(v))^q}, \tag{22}$$

which we will call the q-expectation according to Ref. [58]. Note, that e.g. the exactly solvable case with $\alpha = 1.0$ we find that the ordinary expectation diverges, however as q increases a finite measure is found. Moreover, this gives also the opportunity to define a pseudo-energy in the system as the smallest possible value q where the q-expectation converges. Naturally this reduces to the classical energy for $\alpha = 2$.

In principle, all values of v $F(|v| < \infty)$ should be used for the q-expectation of F(v). However for numerical tractability we have used a PDF with finite support $F(v)_{num} = F(v)$ for $|v| < 10$ and zero everywhere else. Different support ranges have been tested where extending the range $|v| < 15$ makes only minor changes.

It has been proposed analogously to entropy in Boltzmann-Gibbs statistics in thermodynamics that Tsallis entropy may be used in systems where a significant amount of extreme events are present. Entropy is a measure on the number of ways a system can be arranged. In generalized statistical mechanics q-entropy or Tsallis entropy can be introduced as,

$$S_q = \frac{1 - \int dv(F(v))^q}{q-1}. \tag{23}$$

It is interesting to note that the q-entropy is reduced (by L'Hospital's rule) to the conventional Boltzmann-Gibbs entropy $S = -\int dv \ \log(F(v))F(v)$ for Gaussian statistics. The Tsallis entropy will be used to investigate the importance of fractal structure in velocity space and we will contrast the resulting generalized entropies to the standard Boltzmann-Gibbs entropy. The resulting Boltzmann-Gibbs and Tsallis entropies are given in Fig. 4. Contrasting the numerical results with those found by using the numerical modelling, it is found that the Boltzmann-Gibb entropy is increasing with q (indicative of diverging statistical moments) and that the Tsallis entropy is almost flat in the current range and finally seems to decrease for larger q. This is in qualitative agreement with the analytical model however there are some quantitative differences. Note that the range q is smaller compared to the analytical work.

Here in the last part of this paper we will consider the diffusion coefficient using statistical theory. First, investigations of the diffusion coefficient in terms of statistical moments such as skewness and kurtosis of the distribution found earlier in Refs. [59–65], as a first generalization we will find a solution where the velocity derivatives in Eq. 25 is modified to accommodate for fractional transport events in a perturbative manner. Finally we obtain a diffusion equation of the form,

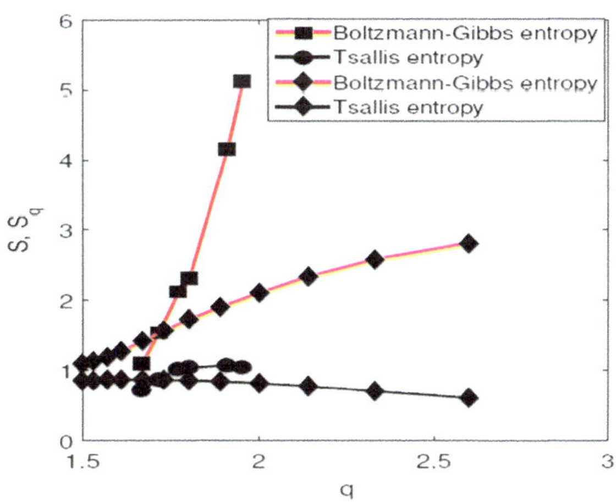

Fig. 4 The numerically (boxes and circles) and analytically (diamonds) found Boltzmann-Gibbs and Tsallis entropies as a function of q using Eq. 20 as a relation between the non-extensivity and fractality

$$\left[\frac{\partial}{\partial t} + \mathbf{v}\cdot\frac{\partial}{\partial \mathbf{r}} - \frac{\partial}{\partial \mathbf{v}}D(\mathbf{v})\frac{\partial}{\partial \mathbf{v}}\right]\langle f_s\rangle = 0, \tag{24}$$

where we evaluate the diffusion by the integral expression,

$$D(\mathbf{v}) = \frac{q_s^2}{m_s^2}\int_0^\infty d\alpha \sum_k \mathbf{E}_k(T+\alpha)\mathbf{E}_{-k}(T)e^{i\mathbf{k}\cdot\mathbf{R}}\langle U\rangle(\alpha)e^{-i\mathbf{k}\cdot\mathbf{R}}. \tag{25}$$

In order to evaluate the expression for the diffusion coefficient in Eq. 25 we restrict ourselves to the weakly nonlinear sector and make use of the cumulant expansion (Gram-Charlier expansion),

$$\langle U(t,0)\exp\left(-i\mathbf{k}\cdot\mathbf{R}\right)\rangle \approx \tag{26}$$

$$\exp\left(-\langle i\mathbf{k}\cdot\mathbf{R}\rangle + \frac{1}{2}\langle i\mathbf{k}\cdot\delta\mathbf{R}(t)\rangle^2 - \langle i\mathbf{k}\cdot\delta\mathbf{R}(t)\rangle^2\rangle + \sum_{n=3}^{\infty}\frac{1}{n}C_n(k,t)\right) \tag{27}$$

Here $\delta\mathbf{R} = \mathbf{R}(t) - \mathbf{R}$ and the coefficients $C_n(k,t)$ are the the third and fourth cumulants are,

$$C_3 = \gamma\sigma^3, \tag{28}$$

$$C_4 = \kappa\sigma^4. \tag{29}$$

Here σ is the standard deviation and the γ and κ are the skewness and kurtosis, respectively.

The average diffusion coefficient is shown in Fig. 5 as a function of the fractality (α). Exponentially decreasing diffusion with increasing fractality in the range [1.25 2.00], here $\alpha = 2.00$ represent the collisionally driven normal diffusion.

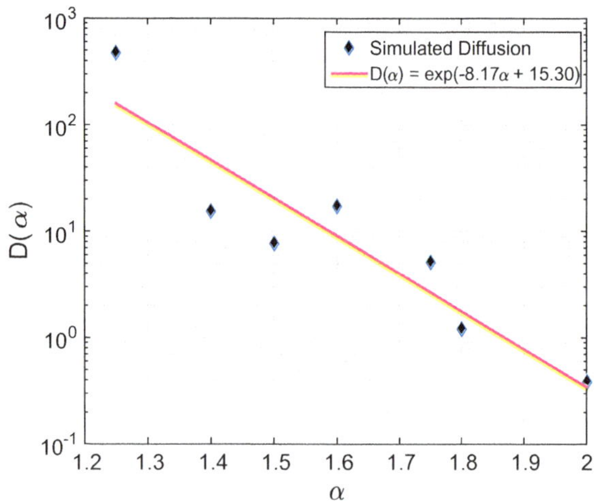

Fig. 5 The numerically estimated diffusion as a function of the fractality index α coefficient based on Langevin model Eqs. (1)–(2)

The effects of electrostatic Lévy type fluctuations is strongly influencing and significantly increases the diffusion for small $\alpha < 2.0$.

4 Summary and Conclusions

Non-linear processes with non-Gaussian character have attracted significant attention during recent years however a consistent and efficient model description is still lacking. In this paper we have investigated two different but connected models based on the Langevin equations with electrostatic fluctuations of Lévy type and fractal Fokker-Planck descriptions.

Monte Carlo numerical simulations of charged particle motion, based on the Langevin equation, in the presence of a fluctuating electric field obeying non-Gaussian Lévy statistics in a constant magnetic field, and a linear friction representing the effects of collisional Coulomb drag have been investigated. The Lévy noise was introduced in order to model the effect of non-local transport due to fractional diffusion in velocity space resulting from intermittent electrostatic turbulence. The statistical properties of the velocity moments and energy for various values of the Lévy index α were investigated, and the role of Lévy fluctuations on the particles Larmor radii, and the statistical moments of displacements were explored. We observed that as α is decreased, the random walk in energy is strongly influenced by outlier events which result in intermittent behaviour with appearance of Lévy flights in between periods of small perturbations. The rate and the amplitude of the intermittent jumps in energy increases significantly as α is decreased. The PDFs of the particles' Larmor radii change from an exponential decay to a power law decay when the stochastic electrostatic process is changed from a Gaussian to a Lévy process. This corroborates the findings in Ref. [36] that the q-moment is an appropriate metric characterizing Lévy distributed processes. Our findings suggest that when turbulent electrostatic fluctuations exhibit non-Gaussian Lévy statistics, gyro-averaging and guiding centre approximations may not be fully justified and full particle orbit effects should be taken into account.

One prominent candidate capturing the main features in the dynamics, namely the Fractional extended Fokker-Planck Equation (FFPE) has been probed theoretically with support of the Monte Carlo simulations. The FFPE is obtained by modifying the velocity derivative to a fractional differential operator allowing for non-local effects in velocity space. The main reason for including the fractional velocity derivative in the FPE is to allow for the non-negligible probability of direction preference and long jumps, i.e., Lévy flights, which therefore allows for asymmetries and long tails in the equilibrium PDFs, respectively. The non-extensive statistical mechanics of Tsallis provides velocity space distribution functions intermediate to that of Gaussians and Lévy distributions adjustable by a continuous real parameter q which seems to be suitable for comparing with the distribution found in FFPE. However, in this case the PDFs are considered to be symmetric. For systems comprised of independent or parts interacting through short-range forces the Boltzmann-Gibb statistical mechanics is sufficient however for systems exhibiting fractal structure or long range correlations this approach becomes unwarranted.

Although some criticism exists on the appropriateness of using the Tsallis statistics in describing processes with Lévy statistics, however, the aim of the present work was to shed light on the non-extensive properties of the velocity space statistics and characterization of the fractal processes of the FFPE in terms of Tsallis statistics in this particular setting. Jespersen et al. [66], showed the example of the Langevin equation with a harmonic potential where the Tsallis q-statistics had limited usefulness. The reason is that using variational calculus of Eq. (23) with the appropriate constraints the relation between α and q is $\alpha = \frac{4-2q}{q-1}$ which is different from Eq. (20) and thus cannot reproduce the correct scalings. They then concluded that the Tsallis entropy was not the appropriate framework for Lévy flights in a harmonic potential described by the generalized Fokker-Planck equation. However, this limitation seems not to impede the usefulness of the application of Tsallis entropy on this Langevin equation where the correct scaling is obtained.

In systems with Lévy type character higher statistical moments diverge however by using the generalized q-moments or q-expectations as,

$$\langle v^p \rangle_q = \int dv F(v)^q v^p / \int dv F(v)^q \tag{30}$$

convergent measures of fractality can be obtained although the regular moments diverges. This also permits us to define a convergent pseudo-energy that is always convergent. In the first step it was shown that the PDFs obtained from the Tsallis statistics could describe the solutions to the FFPE with good agreement. Moreover, we find that self-organising behavior is present in the system where the ratio of the entropy and energy expectation is decreasing with decreasing fractionality or increasing α. Finally, it seems that a FFPE is a viable candidate for explaining certain non-linear features ubiquitous to anomalous plasma transport as well as for other physical processes.

Acknowledgements The author would like to acknowledge contributions and fruitful discussions with Dr. E. Kim and Dr. D. del-Castillo-Negrete.

References

1. M. Schlesinger, G.M. Zaslavsky, J. Klafter, Nature **363**, 31 (1993)
2. I.M. Sokolov, J. Klafter, A. Blumen, Phys. Today **55**, 48 (2002)
3. J. Klafter, I.M. Sokolov, Phys. World **08**, 29 (2005)
4. R. Metzler, J. Klafter Phys. Rep. **339**, 1 (2000)
5. R. Metzler, J. Klafter, J. Phys. A: Math. Gen. **37**, R161 (2004)
6. B.B. Mandelbrot, *The Fractal Geometry of Nature* (W. H. Freeman and Company, San Francisco, 1982)
7. J.A. Krommes, Phys. Rep. **360**, 1–352 (2002)
8. E.W. Montroll, H. Scher, J. Stat. Phys. **9**, 101 (1973)

9. S.C. Kou, X. Sunney Xie, Phys. Rev. Lett. **93**, 180603 (2004)
10. B.A. Carreras, C. Hidalgo, E. Sanchez, M.A. Pedrosa, R. Balbin, I. Garcia-Cortes, B. van Milligen, D.E. Newman, V.E. Lynch, Phys. Plasmas **3**, 2664 (1996)
11. B.A. Carreras, B.P. van Milligen, C. Hidalgo, R. Balbin, E. Sanchez, I. Garcia-Cortes, M.A. Pedrosa, J. Bleuel, E. Endler, Phys. Rev. Lett. **83**, 3653 (1999)
12. B.P. van Milligen, R. Sanchez, B.A. Carreras, V.E. Lynch, B. LaBombard, M.A. Pedrosa, C. Hidalgo, B. Goncalves, R. Balbin, The W7-As Team. Phys. Plasmas **12**, 052507 (2005)
13. R. Sanchez, D.E. Newman, J.-N. Leboeuf, V.K. Decyk, B.A. Carreras, Phys. Rev. Lett. **101**, 205002 (2008)
14. D. del-Castillo-Negrete, B.A. Carreras, V.E. Lynch, Phys. Rev. Lett. **94**, 065003 (2005)
15. J. Anderson, P. Xanthopoulos, Phys. Plasmas **17**, 110702 (2010)
16. E. Kim, H. Liu, J. Anderson, Phys. Plasmas **16**, 052304 (2009)
17. R. Sanchez, B.A. Carreras, D.E. Newman, V.E. Lynch, B.P. van Milligen, Phys. Rev. E **74**, 016305 (2006)
18. S. Moradi, J. Anderson, B. Weyssow, Phys. Plasmas **18**, 062106 (2011)
19. S. Moradi, J. Anderson, Phys. Plasmas **19**, 082307 (2012)
20. S. Moradi, J. Anderson, D. del-Castillo-Negrete, Phys. Plasmas **23**, 090704 (2016)
21. T.S. Hahm, Phys. Fluids **31**, 2670 (1988)
22. S.J. Zweben et al., Plasma Phys. Control. Fusion **49**, S1–S23 (2007)
23. V. Naulin, J. Nucl. Mater. **363–365**, 24–31 (2007)
24. S.M. Kaye et al., Phys. Fluids B **2**, 2926 (1990)
25. N. Lopez Cardozo, Plasma Phys. Controlled Fusion **37**, 799 (1995)
26. K.W. Gentle, R.V. Bravenec, G. Cima, H. Gasquet, G.A. Hallock, P.E. Phillips, D.W. Ross, W.L. Rowan, A.J. Wootton, Phys. Plasmas **2**, 2292 (1995)
27. P. Mantica, P. Galli, G. Gorini, G.M.D. Hogeweij, J. de Kloe, N.J. Lopes Cardozo, Phys. Rev. Lett. **82**, 5048 (1999)
28. B.P. van-Milligen, E. de la Luna, F.L. Tabars, E. Ascasbar, T. Estrada, F. Castejon, J. Castellano, I. Garcia-Cortes, J. Herranz, C. Hidalgo, J.A. Jimenez, F. Medina, M. Ochando, I. Pastor, M.A. Pedrosa, D. Tafalla, L. Garca, R. Sanchez, A. Petrov, K. Sarksian, N. Skvortsova, Nucl. Fusion **42**, 787 (2002)
29. B.A. Carreras et al., Phys. Rev. Lett. **83**, 3653 (1999)
30. G.M. Zaslavsky, Phys. Rep. **371**, 461 (2002)
31. V.E. Tarasov, J. Phys. Conf. Ser. **7**, 17 (2005)
32. V.E. Tarasov, Chaos **16**, 033108 (2006)
33. G.M. Zaslavsky, *Hamiltonian Chaos and Fractional Dynamics* (Oxford University Press, Oxford, 2005)
34. D. del-Castillo-Negrete, B.A. Carreras, V.E. Lynch, Phys. Plasmas **11**, 3854–3864 (2004)
35. D. del-Castillo-Negrete, Nonlinear Process. Geophys. **17**, 795–807 (2010)
36. J. Anderson, E. Kim, S. Moradi, Phys. Plasmas **21**, 122109 (2014)
37. A.V. Chechkin, V. Yu, Gonchar, M. Szydlowski, Phys. Plasmas **9**, 78 (2002)
38. S. Borak, W. Härdle, R. Weron, *Statistical Tools for Finance and Insurance* (Springer, Berlin, 2005), pp. 21–44. http://ideas.repec.org/p/hum/wpaper/sfb649dp2005-008.html
39. J.M. Chambers, C.L. Mallows, B.W. Stuck, A method for simulating stable random variables. JASA **71**, 340–344 (1976)
40. R. Weron, J.E. Gentle, W. Härdle, Y. Mori, *Computationally intensive value at risk calculations, in Handbook of Computational Statistics: Concepts and Methods* (Springer, Berlin, 2004), pp. 911–950
41. I. Pavlyukevich, M. Riedle, Stoch. Anal. Appl. **33**, 271 (2015)
42. B. Weyssow, A. Vulpiani, F. Mainardi, et al., *Anomalous Transport: Foundations and Applications*, ed. by R. Klages, G. Radons, I.M. Sokolov (Wiley, Germany, 2008)
43. D. del-Castillo-Negrete, Non-diffusive transport in fusion plasmas: fractional diffusion approach, in *Proceedings of the "First ITER Summer School: Turbulent Transport in Fusion Plasmas." AIP Conference Proceedings*, vol. 1013 (AIP, Melville, New York, 2008)
44. S. Chandrasekhar, Rev. Mod. Phys. **21**, 383 (1949)

45. A.Y. Khintchine, *The Mathematical Foundation of Statistical Mechanics* (Dover, New York, 1948)
46. P. Lévy *Theorie del 'Addition des Variables* (Gauthier-Villiers, Paris, 1937)
47. B.J. West, V. Seshadri, Phys. A **113**, 203 (1982)
48. H.C. Fogedby, Phys. Rev. E **50**, 1657 (1994)
49. H.C. Fogedby, Phys. Rev. Lett. **73**, 2517 (1994)
50. E. Barkai, Phys. Rev. E. Rapid Commun. **68**, 055104(R) (2003)
51. C. Tsallis, A.M.C. de Souza, R. Maynard, Lect. Notes Phys. **450**, 269 (1995)
52. C. Tsallis, D.J. Bukman, Phys. Rev. E **54**, R2197 (1996)
53. C. Tsallis, R.S. Mendes, A.R. Plastino, Phys. A **261**, 534 (1998)
54. G. Balasis, I.A. Daglis, A. Anastasiadis, C. Papadimitriou, M. Mandea, K. Eftaxias, Phys. A **390**, 341 (2011)
55. G.P. Pavlos, L.P. Karkatsanis, M.N. Xenakis, D. Sarafopoulos, E.G. Pavlos, Phys. A **391**, 3069 (2012)
56. G.P. Pavlos, L.P. Karkatsanis, M.N. Xenakis, Phys. A **391**, 6287 (2012)
57. C. Tsallis, S.V.F. Lévy, A.M.C. Souza, R. Maynard, Phys. Rev. Lett. **75**, 3589 (1995)
58. D. Prato, C. Tsallis, Phys. Rev. E **60**, 2398 (1999)
59. B.B. Kadomtsev, *Plasma Turbulence*, Chap. III (Academic Press Inc., New York, 1965)
60. T.H. Dupree, Phys. Fluids **9**, 1773 (1966)
61. T.H. Dupree, Phys. Fluids **10**, 1049 (1967)
62. S.A. Orszag, R.H. Kraichnan, Phys. Fluids **10**, 1720 (1967)
63. A.I. Saichev, G.M. Zaslavsky, Chaos **7**, 753 (1997)
64. J. Weinstock, Phys. Fluids **12**, 1045 (1969)
65. J. Weinstock, Phys. Fluids **13**, 2308 (1970)
66. S. Jespersen, R. Metzler, H.C. Fogedby, Phys. Rev. E **59**, 2736 (1999)

Analysis of Low-Frequency Instabilities in Low-Temperature Magnetized Plasma

Dan-Gheorghe Dimitriu and Maricel Agop

1 Introduction

The study of the plasma instabilities is still a hot topic in physics, especially because of their involvement in the physics of controlled nuclear fusion, as well as in maintaining stable regimes for technological plasmas [31]. In low-temperature magnetized plasmas, most of the experimental investigations of instabilities were carried out in the Q-machines [4–7, 13, 14, 19, 20, 22, 30], the most investigated instabilities being the potential relaxation instability (PRI) and the electrostatic ion-cyclotron instability (EICI). These instabilities appear as oscillations of the plasma parameters, sometimes showing strongly nonlinear behavior.

PRI and EICI are excited by drawing an electron current parallel to the magnetic field to a disk electrode which is inserted into the plasma column perpendicular to the axis. For exciting the PRI, the radius of the electrode has to be larger than the plasma column, while for exciting the EICI, the radius of the electrode must be considerably smaller than that of the plasma column, but still in range of a few ion gyroradii. A transition from PRI to EICI by decreasing the diameter of the electrode was reported by Schrittwieser [35]. Dimitriu et al. [13] reported on the simultaneous excitation of PRI and EICI, with comparable amplitudes. This led to a strong modulation of the EICI by PRI, which affected not only the amplitude but also the frequency of the EICI. Because of this, in the spectrum of the current oscillations sidebands around f_{EICI} appear with a frequency difference equal to $\pm f_{\mathrm{PRI}}$.

D.-G. Dimitriu (✉)
Faculty of Physics, "Alexandru Ioan Cuza" University, Iasi, Romania
e-mail: dimitriu@uaic.ro

M. Agop
Department of Physics, Faculty of Machine Manufacturing and Industrial Management, "Gheorghe Asachi" Technical University, Iasi, Romania
e-mail: m.agop@yahoo.com

C. H. Skiadas (ed.), *Fractional Dynamics, Anomalous Transport and Plasma Science*, https://doi.org/10.1007/978-3-030-04483-1_5

Traditionally, the magnetized plasma of a Q-machine was considered to be collisionless. Starting with this assumption, phenomenological models were developed for PRI [19] and EICI [34]. However, Sanduloviciu [32] proposed a new model for EICI, by supposing that this instability is triggered by the dynamics of a 3D double layer developed in front of the electrode as the result of the electron-neutral impacts excitation and ionization, similar to the phenomenology of the anodic fireballs [33]. The same assumption of low collisionality in the Q-machine plasma was considered by Avram et al. [4–7] to explain the nonlinear effects experimentally observed in relation to PRI and EICI. Dimitriu et al. [13] explained the interaction between the simultaneous excited PRI and EICI also by taking into consideration of collisions in the Q-machine plasma, concluding that the traditional opinion that a Q-machine plasma can be considered collisionless cannot be maintained.

To develop a new model we have to admit that the Q-machine plasma can achieve self-similarity (space-time structures can appear) associated with strong fluctuations at all possible space-time scales [11, 12, 15]. Then, for time scales that prove to be larger when compared with the inverse of the highest Lyapunov exponent, the deterministic trajectories are replaced by a collection of potential routes. At its turn, the concept of "definite position" is replaced by that of probability density [24, 28, 29]. The most impressive example in this respect refers to collision processes in Q-machine plasma. By accepting the presence of collisions in the Q-machine plasma, the dynamics of the particles can be described through non-differentiable curves (i.e. fractal curves). Then, the fractality appears as a universal property of the Q-machine plasma and the complexity of interactions in the dynamics of the Q-machine plasma is replaced by fractality.

Since the theoretical models which describe the dynamics of the Q-machine plasma are sophisticated, in the present paper a new model will be developed using the Scale Relativity Theory (SRT) with arbitrary constant fractal dimension [1–3, 8–10, 16, 17, 25–27]. The model is able to predict the interaction between the simultaneously excited PRI and EICI, phenomenon which is experimentally evidenced.

2 Hallmarks of Fractality

We can simplify the dynamics of the Q-machine plasma by assuming that its particles (electrons, ions, neutrals) move on fractal curves. Once accepted such a hypothesis, the dynamics of the Q-machine plasma particles is given by the fractal operator $\hat{\mathrm{d}}/\mathrm{d}t$ [1, 10]:

$$\frac{\hat{\mathrm{d}}}{\mathrm{d}t} = \frac{\partial}{\partial t} + \hat{\boldsymbol{V}} \cdot \nabla - i\frac{\lambda^2}{\tau}\left(\frac{\mathrm{d}t}{\tau}\right)^{(2/D_F)-1}\Delta, \tag{1}$$

where

$$\hat{\boldsymbol{V}} = \boldsymbol{V}_D - i\boldsymbol{V}_F \tag{2}$$

is the complex velocity, $\boldsymbol{V_D}$ is the differentiable and resolution scale independent velocity, $\boldsymbol{V_F}$ is the non-differentiable and resolution scale dependent velocity. In Eq. (1), $\hat{\boldsymbol{V}} \cdot \nabla$ is the convective term, $\frac{\lambda^2}{\tau}\left(\frac{dt}{\tau}\right)^{(2/D_F)-1}\Delta$ is the dissipative term, D_F is the fractal dimension of the movement curve, λ is the space scale, τ is the time scale, dt/τ is the scale resolution and λ^2/τ is a specific coefficient associated to the fractal—non-fractal transition. For D_F any definition can be used (the Hausdorff-Besikovici fractal dimension, the Kolmogorov fractal dimension, etc. [24]), but once accepted such a definition for D_F, it has to be maintained over the entire analysis. In a particular case, when the motions of the Q-machine plasma particles are considered on Peano curves, i.e. $D_F = 2$, the fractal operator (1) reduces to Nottale's operator [28, 29].

Applying the fractal operator (1) to the complex velocity (2) and accepting the principle of scale covariance [28, 29] in the form

$$\frac{\hat{d}\hat{\boldsymbol{V}}}{dt} = 0, \tag{3}$$

the motion equation is obtained:

$$\frac{\hat{d}\hat{\boldsymbol{V}}}{dt} = \frac{\partial\hat{\boldsymbol{V}}}{\partial t} + \left(\hat{\boldsymbol{V}} \cdot \nabla\right)\hat{\boldsymbol{V}} - i\frac{\lambda^2}{\tau}\left(\frac{dt}{\tau}\right)^{(2/D_F)-1}\Delta\hat{\boldsymbol{V}} = 0. \tag{4}$$

It means that at any point of a fractal path, the local acceleration term, $\partial_t\hat{\boldsymbol{V}}$, the non-linearly (convective) term, $\left(\hat{\boldsymbol{V}} \cdot \nabla\right)\hat{\boldsymbol{V}}$ and the dissipative term, $\left(\lambda^2/\tau\right)(dt/\tau)^{(2/D_F)-1}\Delta\hat{\boldsymbol{V}}$, make their balance. Therefore, the Q-machine plasma dynamics can be assimilated with the dynamics of a "rheological" fluid. This dynamics is described by the complex velocity field $\hat{\boldsymbol{V}}$, the complex acceleration field $\hat{d}\hat{\boldsymbol{V}}/dt = 0$ and by the imaginary viscosity type coefficient $i\left(\lambda^2/\tau\right)(dt/\tau)^{(2/D_F)-1}$.

For irrotational motions of the Q-machine plasma particles

$$\nabla \times \hat{\boldsymbol{V}} = 0, \quad \nabla \times \boldsymbol{V_D} = 0, \quad \nabla \times \boldsymbol{V_F} = 0, \tag{5}$$

$\hat{\boldsymbol{V}}$ can be chosen on the form:

$$\hat{\boldsymbol{V}} = -i\frac{\lambda^2}{\tau}\left(\frac{dt}{\tau}\right)^{(2/D_F)-1}\nabla \ln\psi, \tag{6}$$

where $\phi = \ln\psi$ is the velocity scalar potential. By substituting (6) in (4) and using the method described by Munceleanu et al. [25], it results:

$$\frac{\hat{d}\hat{\boldsymbol{V}}}{dt} = -i\frac{\lambda^2}{\tau}\left(\frac{dt}{\tau}\right)^{(2/D_F)-1}\nabla\left[\frac{\partial \ln\psi}{\partial t} - i\frac{\lambda^2}{\tau}\left(\frac{dt}{\tau}\right)^{(2/D_F)-1}\frac{\nabla\psi}{\psi}\right] = 0. \tag{7}$$

This equation can be integrated in a universal way and yields

$$\frac{\lambda^4}{\tau^2}\left(\frac{\mathrm{d}t}{\tau}\right)^{(4/D_F)-2}\Delta\psi + i\frac{\lambda^2}{\tau}\left(\frac{\mathrm{d}t}{\tau}\right)^{(2/D_F)-1}\frac{\partial\psi}{\partial t} = 0, \tag{8a}$$

up to an arbitrary phase factor which may be set to zero by a suitable choice of the phase of ψ. The presence of an external scalar potential φ transforms Eq. (8a) in the form:

$$\frac{\lambda^4}{\tau^2}\left(\frac{\mathrm{d}t}{\tau}\right)^{(4/D_F)-2}\Delta\psi + i\frac{\lambda^2}{\tau}\left(\frac{\mathrm{d}t}{\tau}\right)^{(2/D_F)-1}\frac{\partial\psi}{\partial t} - \frac{\varphi}{2}\psi = 0. \tag{8b}$$

For motions of plasma particles on Peano's curves, $D_F = 2$, the Eqs. (8a, 8b) takes the Nottale's form [28, 29]. Moreover, for motions of plasma particles on Peano's curves at Compton scale, $\lambda^2/\tau = \hbar/2\mu$, with $\hbar$ the reduced Planck constant and μ the reduced mass of the plasma particles, the relation (8a) becomes the standard Schrödinger equation. Such a physical situation can be found in the case of degenerate plasmas (fusion hot plasmas).

If $\psi = \sqrt{\rho}e^{iS}$ with $\sqrt{\rho}$ the amplitude and S the phase of ψ, the complex velocity field (6) takes the explicit form:

$$\hat{\boldsymbol{V}} = \frac{\lambda^2}{\tau}\left(\frac{\mathrm{d}t}{\tau}\right)^{(2/D_F)-1}\nabla S - i\frac{\lambda^2}{2\tau}\left(\frac{\mathrm{d}t}{\tau}\right)^{(2/D_F)-1}\nabla \ln\rho, \tag{9a}$$

$$\boldsymbol{V_D} = \frac{\lambda^2}{\tau}\left(\frac{\mathrm{d}t}{\tau}\right)^{(2/D_F)-1}\nabla S, \tag{9b}$$

$$\boldsymbol{V_F} = \frac{\lambda^2}{2\tau}\left(\frac{\mathrm{d}t}{\tau}\right)^{(2/D_F)-1}\nabla \ln\rho. \tag{9c}$$

By substituting (9a–9c) in (4) and separating the real and the imaginary parts, up to an arbitrary phase factor which may be set at zero by a suitable choice of the phase of ψ, we obtain:

$$\frac{\partial \boldsymbol{V_D}}{\partial t} + (\boldsymbol{V_D}\cdot\nabla)\boldsymbol{V_D} = -\nabla Q, \tag{10a}$$

$$\frac{\partial\rho}{\partial t} + \nabla\cdot(\rho\boldsymbol{V_D}) = 0, \tag{10b}$$

with Q the specific fractal potential

$$Q = -2\frac{\lambda^4}{\tau^2}\left(\frac{\mathrm{d}t}{\tau}\right)^{(4/D_F)-2}\frac{\Delta\sqrt{\rho}}{\sqrt{\rho}} = -\frac{\boldsymbol{V_F^2}}{2} - \frac{\lambda^2}{\tau}\left(\frac{\mathrm{d}t}{\tau}\right)^{(2/D_F)-1}\nabla\cdot\boldsymbol{V_F}. \tag{11}$$

Equation (10a) represents the specific momentum conservation law, while Eq. (10b) represents the states density conservation law. Equations (10a–10b) and (11) define the fractal hydrodynamics model (FHM).

The following conclusions are obvious:

(i) any particle of the Q-machine plasma is in permanent interaction with the fractal medium through the specific fractal potential (11);
(ii) the fractal medium is identified with a non-relativistic fractal fluid described by the specific momentum and probability density conservation laws, (10a-10b) (FHM);
(iii) the fractal speed $\boldsymbol{V_F}$ does not represent an actual mechanical model, but contribute to the transfer of specific momentum and the concentration of energy. This may be clearly seen from the absence of $\boldsymbol{V_F}$ from the probability density conservation law (10b) and from its role in the variational principle [28, 29];
(iv) any interpretation of the fractal potential Q should take cognizance of the "self" or internal nature of the specific momentum transfer. While the energy is stored in the form of the mass motion and potential energy (as classically it is), some is available elsewhere and only the total one is conserved. It is the conservation of energy and specific momentum that ensures the reversibility and existence of eigenstates, but denies a Brownian motion type form of interaction with an external medium;
(v) the specific fractal potential must be considered as a kinetic term and not as a potential one. This generates the viscosity stress tensor type [25]:

$$\hat{\sigma}_{il} = \frac{\lambda^4}{\tau^2}\left(\frac{\mathrm{d}t}{\tau}\right)^{(4/D_F)-2}\left(\nabla_i\nabla_l\rho - \frac{\nabla_i\rho\nabla_l\rho}{\rho}\right) = \eta\left(\frac{\partial V_{Fl}}{\partial x_i} + \frac{\partial V_{Fi}}{\partial x_l}\right), \quad (12)$$

with $\eta = (\rho/2)\left(\lambda^2/\tau\right)(\mathrm{d}t/\tau)^{(2/D_F)-1}$ a viscosity type coefficient, which divergence is equal to the usual force density associated to Q:

$$\nabla_i\hat{\sigma}_{il} = -\rho\nabla_l Q. \quad (13)$$

The diagonal form of the viscosity stress type tensor, $\hat{\sigma}_{il} = \sigma\delta_{il}$, where δ_{il} is the Kronecker's pseudo-tensor, implies correspondences of FHM with the standard hydrodynamic model [23];
(vi) since the position vector of the Q-machine plasma particle is assimilated to a stochastic process of Wiener type [24, 28, 29], ψ is not only the scalar potential of complex velocity (through $\ln\psi$) in the fractal hydrodynamics, but also the density of probability (through $|\psi|^2$) in the Schrödinger type theory. It results the equivalence between the formalism of the fractal hydrodynamics and the Schrödinger type one. Moreover, the chaoticity, either through turbulence in the fractal hydrodynamics approach or through stochasticization in the Schrödinger type approach, is generated only by the non-differentiabilty of the movement trajectories in a fractal space.

3 Potential Relaxation Instability

The phenomenology of the PRI was described [12–14] by considering that, after a self-organization process, in front of the Q-machine cold plate a double layer appears, of which dynamics determines the dynamics of another double layer between the Q-machine hot and cold plates. Thus, in this case, the dynamics reduces to that one of an interface generated at the contact of two different plasmas, from which one is enriched in positive ions. Mathematically, such a dynamics can be described by using the Schrödinger type Eq. (8b) in the form:

$$2i\mu D\frac{d\psi}{dt} = H\psi, \tag{14a}$$

$$D = \frac{\lambda^2}{\tau}\left(\frac{dt}{\tau}\right)^{(2/D_F)-1}, \tag{14b}$$

applied to the initially isolated and finally in contact plasmas, where H is the associated "Hamiltonian".

Then, let's consider two physical objects (associated to the two plasmas) separated by an interface (associated to the double layer). If the interface is thick enough so that the physical objects are isolated from each other, the time-dependent Schrödinger type equations [see Eq. (8b)] for each side are:

$$2i\mu D\frac{d\psi_1}{dt} = H_1\psi_1, \tag{15a}$$

$$2i\mu D\frac{d\psi_2}{dt} = H_2\psi_2, \tag{15b}$$

where ψ_i are the wave functions of the physical objects and H_i are the "Hamiltonians" on either side of the interface. Let's assume that a voltage $2U$ is applied between the two physical objects. If the zero value of the potential is assumed to be in the middle of the interface, let's consider that the potential of the physical object 1 is $-U$, while the potential of the physical object 2 is $+U$.

The interface presence leads to a coupling of the Eqs. (15a, 15b) in the form

$$2\mu D\frac{d\psi_1}{dt} = eU\psi_1 + \Gamma\psi_2, \tag{16a}$$

$$2\mu D\frac{d\psi_2}{dt} = -eU\psi_2 + \Gamma\psi_1, \tag{16b}$$

where Γ is the coupling constant for the wave functions across the interface. Since the square of each wave function is a probability density, the two wave functions can be written in the form:

$$\psi_1 = \sqrt{\rho_1}e^{i\theta_1}, \tag{17a}$$

$$\psi_1 = \sqrt{\rho_1}e^{i\theta_1}, \tag{17b}$$

$$\Theta = \theta_2 - \theta_1. \tag{17c}$$

If the two wave functions are substituted in the coupled Eqs. (16a, 16b) and the results are separated into real and imaginary parts, the time dependences of the particle densities and the phase difference are obtained:

$$\frac{\mathrm{d}\rho_1}{\mathrm{d}t} = \frac{\Gamma}{\mu D}\sqrt{\rho_1\rho_2}\sin\Theta, \tag{18a}$$

$$\frac{\mathrm{d}\rho_2}{\mathrm{d}t} = -\frac{\Gamma}{\mu D}\sqrt{\rho_1\rho_2}\sin\Theta, \tag{18b}$$

$$\frac{\mathrm{d}\Theta}{\mathrm{d}t} = \frac{eV}{\mu D} = \Omega. \tag{18c}$$

The current density can be specified in terms of the difference between Eqs. (18a) and (18b), multiplied by e

$$j = e\frac{\mathrm{d}}{\mathrm{d}t}(\rho_1 - \rho_2) = j_c \sin\Theta, \tag{19}$$

where

$$j_c = \frac{2e\Gamma}{\mu D}\sqrt{\rho_1\rho_2} \tag{20}$$

is the critical current.

From (18c) it results that a phase change accompanies the presence of a voltage across a interface. Since the applied voltage is constant, (18c) can be directly integrated, giving

$$\Theta(t) = \Theta_0 + \frac{eU}{m_0 D}t - \Theta + \Omega t, \tag{21a}$$

$$\Theta_0 = \text{const.}, \tag{21b}$$

which provides the characteristic frequency (18c). Now, the current density can be written in the form:

$$j = j_c \sin(\Omega t + \Theta_0). \tag{22}$$

Thus, PRI behaves as periodic oscillations of the current (22) collected by the Q-machine cold plate, with the characteristic frequency (18c). By choosing the fractal—non-fractal diffusion coefficient on the form $D = dc$, where d is the distance between the hot and cold plates and c is the ion-acoustic velocity, the characteristic frequency takes the form:

$$\Omega = \frac{eU}{2\mu dc}, \tag{23}$$

being proportional with the applied voltage on the cold plate and inverse proportional with the distance between the Q-machine hot and cold plates. Such proportionality was experimentally observed [13, 19].

4 Electrostatic Ion-Cyclotron Instability

EICI behaves as coherent oscillations of the Q-machine plasma potential with a frequency slightly higher than the ion-cyclotron frequency. The electron gyroradius is much smaller than the ion gyroradius, so that it can be considered that the electrons don't move in the direction perpendicular to the magnetic field lines. So, the electrons can be considered as inertials, with Boltzmann distribution.

Let's reconsider the equations of FHM for ionic fluids, the electronic one being with Boltzmann distribution:

$$n_e = n_0 \exp\left(\frac{eU}{k_B T_e}\right), \tag{24}$$

where n_0 is the unperturbed plasma density, U is the voltage, T_e is the electron temperature and k_B is the Boltzmann constant. According to (10a, 10b) of the FHM and choosing the diagonal form of $\hat{\sigma}_{il}$, the ion dynamics can be characterized by:

(a) one-dimensional motion equation

$$m_i\left(\frac{\partial V_i}{\partial t} + V_i \frac{\partial V_i}{\partial x}\right) = -q\frac{\partial U}{\partial x} + qV_iB, \tag{25}$$

(b) one-dimensional continuity equation

$$\frac{\partial n_i}{\partial t} + \frac{\partial (n_i V_i)}{\partial x} = 0, \tag{26}$$

together with the one-dimensional Poisson's equation

$$\frac{\partial^2 U}{\partial x^2} = \frac{e}{\varepsilon_0}(n_e - n_i), \tag{27}$$

where V_i is the ion velocity, n_i is the ions density, T_i is the ion temperature, m_i is the ion inertial mass, x and t are the spatial and temporal coordinates, q is the electric charge, B is the magnetic field induction and ε_0 is the vacuum permittivity. Now, by introducing the dimensionless variables

$$\phi = \frac{eV}{kT_e}, \tag{28a}$$

$$V = \frac{V_i}{c}, \tag{28b}$$

$$n = \frac{n_i}{n_0}, \tag{28c}$$

$$\xi = \frac{x}{\lambda}, \tag{28d}$$

$$\tau = \omega t, \tag{28e}$$

where λ is a characteristic length, ω is a characteristic frequency and $c = (kT_e/m_i)^{1/2}$ is the ion-acoustic speed, the Eqs. (25), (26) and (27) become

$$\frac{\partial V}{\partial \tau} + V\frac{\partial V}{\partial \xi} = -\frac{\partial \phi}{\partial \xi} + \frac{\omega_B}{\omega}V, \tag{29a}$$

$$\frac{\partial n}{\partial \tau} + \frac{\partial}{\partial \xi}(nV) = 0, \tag{29b}$$

$$\frac{\partial^2 \phi}{\partial \xi^2} = \left(e^{\phi} - n\right), \tag{29c}$$

where $\omega_B = qB/m_i$. A stationary solution of the equations system (29a–29c) is searched, by introducing the spatio-temporal dependence through the variable

$$\eta = \xi - M\tau, \tag{30}$$

where $M = w/c$ is the Mach's number and w is the double layer velocity. By taking into account (30), the equations system (29a–29c) become

$$-M\frac{\mathrm{d}V}{\mathrm{d}\eta} + V\frac{\mathrm{d}V}{\mathrm{d}\eta} = -\frac{\mathrm{d}\phi}{\mathrm{d}\eta} + \frac{\omega_B}{\omega}V, \tag{31a}$$

$$-M\frac{\mathrm{d}n}{\mathrm{d}\eta} + \frac{\mathrm{d}}{\mathrm{d}\eta}(nV) = 0, \tag{31b}$$

$$\frac{\mathrm{d}^2\phi}{\mathrm{d}\eta^2} = \left(e^{\phi} - n\right), \tag{31c}$$

where further, by integrating the relations (31a, 31b), it follows

$$-MV + \frac{V^2}{2} = -\phi + \frac{\omega_B}{\omega}\int V(\eta)\mathrm{d}\eta + C_1, \tag{32a}$$

$$-Mn + nV = C_2, \tag{32b}$$

C_1 and C_2 being two integration constants. By imposing the limit conditions:

$$\phi(\eta \to \pm\infty) \to 0, \tag{33a}$$

$$V(\eta \to \pm\infty) = 0, \tag{33b}$$

$$n(\eta \to \pm\infty) \to 1, \tag{33c}$$

and resolving the integral from (32a) by the method described by Jackson [21] and Hou [18], eliminating the velocity between (32a) and (32b), and taking into consideration just the first order terms in ϕ, the following relation is obtained:

$$n \approx 1 - \frac{\phi}{\overline{M}^2}, \tag{34a}$$

$$\overline{M}^2 \approx 1 + 1.42\left(\frac{\omega_B}{\omega}\right)^2. \tag{34b}$$

In these conditions, the Poisson equation (31c) takes the form (in the same approximation for ϕ as above):

$$\frac{d^2\phi}{d\eta^2} \approx \left(1 - \frac{1}{\overline{M}^2}\right)\phi. \tag{35}$$

In particular, for ξ = const., Eq. (35) in temporal coordinate becomes:

$$\frac{d^2U}{dt^2} + 1.4\omega_B^2 U \approx 0, \tag{36}$$

which specifies the periodic potential

$$U = U_0 \cos(1.19\omega_B t + \varphi_0), \quad \varphi_0 = \text{const.} \tag{37}$$

Thus, an oscillatory potential is obtained, with the frequency $\Omega_B = 1.19\omega_B$, in agreement with the experimental findings [30].

5 Interaction Between Potential Relaxation Instability and Electrostatic Ion-Cyclotron Instability

In the above conditions, if one overlays the ac voltage (37) over the dc voltage from (22), the current density can be written as an infinite series of products of Bessel functions J_n and sine waves

$$\begin{aligned} j(t) &= j_c \sin\left(\frac{eU}{\mu D}t + \frac{eU_0}{\mu D}t\cos\Omega t + \Theta_0\right) \\ &= j_c \sum_n (-1)^n J_n\left(\frac{eU_0}{\mu D\Omega}\right)\sin\left[(\Omega_B - n\Omega)t + \Theta'\right] \end{aligned} \tag{38}$$

where Θ' is an integration constant. Thus, the amplitude and frequency modulations of the current density are obtained.

6 Experimental Confirmation of the Interaction Between Potential Relaxation Instability and Electrostatic Ion-Cyclotron Instability

The experiments were performed in the single-ended Q-machine of the University of Innsbruck, extensively described in other works [13, 14]. The background pressure of the residual gases was less than 10^{-5} mbar. The potassium plasma density was $n_{\mathrm{pl}} \cong 10^8$–$10^9$ cm^{-3} and the ion and electron temperature were $T_e \cong T_i \cong 0.2$ eV. The confining magnetic field was 0.05 T$<B<$0.2 T, while the diameter of the plasma column was of 3.5 cm.

At a distance of $d = 27.5$ cm from the hot plate a tantalum disk electrode (cold plate) of 1 cm diameter was inserted in the center of the plasma column, perpendicular to the magnetic field lines. By positively biasing the cold plate, a current was drawn through a channel of roughly the same diameter as the cold plate. By increasing the voltage applied on the cold plate, the EICI is the first instability which appears, with a frequency of about $f_{\mathrm{EICI}} \cong 67$ kHz. The identity of the EICI has been proved by measuring the well-known linear increase of the frequency with the magnetic field induction B. The PRI appears at higher values of the voltage, with small amplitude and a frequency of about $f_{\mathrm{PRI}} \cong 15$ kHz. Further increasing the voltage applied on the cold plate leads to an increase of the PRI amplitude. When the amplitudes of the two instabilities become of almost the same order of magnitude, the interaction between the instabilities leads to a strong modulation of the EICI by the PRI, not only in amplitude (see the oscillations of the current collected by the cold plate in Fig. 1a) but also in frequency (see the Fast Fourier Transform of the current oscillations collected by the cold plate in Fig. 1b), and consequently sidebands around f_{EICI} are formed with a frequency $f_{\mathrm{EICI}} \pm f_{\mathrm{PRI}}$. The EICI amplitude modulation with the PRI frequency affects more strongly the negative excursions of the current (see Fig. 1a) because of the current limitation due to the formation of thermal barriers in front of the periodically traveling double layer, a well-known feature of both instabilities [19, 30, 34].

The experimental results shown in Fig. 1a, b confirm the prediction of the theoretical model through the relation (38).

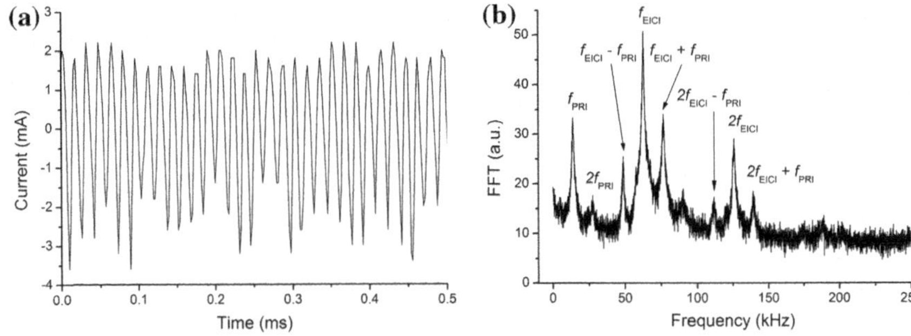

Fig. 1 Time series of the current oscillations collected by the electrode **a** and their FFT **b**, showing the strong interaction between the PRI and EICI

7 Conclusions

A theoretical model was developed in the frame of the scale relativity theory to describe some characteristics of two low-frequency instabilities of the magnetized plasma produce by a Q-machine: potential relaxation instability and electrostatic ion-cyclotron instability. It was found that the potential relaxation instability frequency is proportional with the voltage applied on the cold plate and inverse proportional with the distance between the cold and hot plates of the Q-machine, as experimentally observed. In the case of electrostatic ion-cyclotron instability, its frequency was found to be 1.19 times ion-cyclotron frequency, a result also in excellent agreement with the experimental findings.

The theoretical model is also able to explain interaction between the two above-mentioned instabilities, which leads to the amplitude and frequency modulation of the second instability by the first one. Experimental result are shown, which are in agreement with the theoretical model predictions.

Acknowledgements This work was supported by a grant of Romanian Ministry of Research and Innovation, CNCS—UEFISCDI, project number PN-III-P4-ID-PCE-2016-355, within PNCDI III.

References

1. M. Agop, N. Forna, I. Casian-Botez, C. Bejenariu, New theoretical approach of the physical processes in nanostructures. J. Comput. Theor. Nanosci. **5**, 483–489 (2008)
2. M. Agop, P. Nica, M. Girtu, On the vacuum status in Weyl-Dirac theory. Gen. Relativ. Gravit. **40**, 35–55 (2008)
3. M. Agop, P. Nica, O. Niculescu, D.G. Dimitriu, Experimental and theoretical investigations of the negative differential resistance in a discharge plasma. J Phys Soc Japan **81**, 064502 (2012)
4. C. Avram, R. Schrittwieser, M. Sanduloviciu, Possible excitation and ionization processes in a “collisionless” alkaline plasma. Int. J. Mass Spectrom. **184**, 129–143 (1999)

5. C. Avram, R. Schrittwieser, M. Sanduloviciu, Current jumps and hysteresis in a single-ended Q-machine in connection with the electrostatic ion-cyclotron instability. Contrib. Plasma Phys. **39**, 223–233 (1999)
6. C. Avram, R. Schrittwieser, M. Sanduloviciu, Nonlinear effects in the current-voltage characteristic of a low-density Q-machine plasma: I. Related to the potential relaxation instability. J. Phys. D Appl. Phys. **32**, 2750–2757 (1999)
7. C. Avram, R. Schrittwieser, M. Sanduloviciu, Nonlinear effects in the current-voltage characteristic of a low-density Q-machine plasma: II. Related to the electrostatic ion-cyclotron instability. J. Phys. D: Appl. Phys. **32**, 2758–2762 (1999)
8. I. Casian-Botez, M. Agop, P. Nica, V. Paun, V. Munceleanu, Conductive and convective types behaviors at nano-time scales. J. Comput. Theor. Nanosci. **7**, 2271–2280 (2010)
9. C. Ciubotariu, M. Agop, Absence of a gravitational analog to the Meissner effect. Gen. Relativ. Gravit. **28**, 405–412 (1996)
10. M. Colotin, G.O. Pompilian, P. Nica, S. Gurlui, V. Paun, M. Agop, Fractal transport phenomena through the scale relativity model. Acta Phys. Pol. A **116**, 157–164 (2009)
11. N. D'Angelo, R.W. Motley, Electrostatic oscillations near the ion cyclotron frequency. Phys. Fluids **5**, 633–634 (1962)
12. D.G. Dimitriu, *Electrostatic Instabilities in the Q-machine Plasma (in Romanian)* (Demiurg Editorial House, Iasi, Romania, 2006)
13. D.G. Dimitriu, V. Ignatescu, C. Ionita, E. Lozneanu, M. Sanduloviciu, R.W. Schrittwieser, The influence of electron impact ionisations on low frequency instabilities in a magnetised plasma. Int. J. Mass Spectrom. **223–224**, 141–158 (2003)
14. D.G. Dimitriu, C. Ionita, R. Schrittwieser, Nonlinear effects related to the simultaneous excitation of three instabilities in magnetized plasma. Contrib. Plasma Phys. **51**, 554–559 (2011)
15. W.E. Drummond, M.N. Rosenbluth, Anomalous diffusion arising from microinstabilities in a plasma. Phys. Fluids **5**, 1507–1513 (1962)
16. S. Gurlui, M. Agop, M. Strat, S. Bacaita, Some experimental and theoretical results on the anodic patterns in plasma discharge. Phys. Plasmas **13**, 063503 (2006)
17. S. Gurlui, M. Agop, P. Nica, M. Ziskind, C. Focsa, Experimental and theoretical investigations of transitory phenomena in high-fluence laser ablation plasma. Phys. Rev. E **78**, 026405 (2008)
18. T.Y. Hou, *Multi-Scale Phenomena In Complex Fluids: Modeling, Analysis and Numerical Simulations* (World Scientific, Singapore, 2009)
19. S. Iizuka, P. Michelsen, J.J. Rasmussen, R. Schrittwieser, R. Hatakeyama, K. Saeki, N. Sato, Dynamics of a potential barrier formed on the tail of a moving double layer in a collisionless plasma. Phys. Rev. Lett. **48**, 145–148 (1982)
20. S. Iizuka, P. Michelsen, J.J. Rasmussen, R. Schrittwieser, R. Hatakeyama, K. Saeki, N. Sato, Double layer dynamics in a collisionless magnetoplasma. J. Phys. Soc. Jpn. **54**, 2516–2529 (1985)
21. E.A. Jackson, in *Perspectives in Nonlinear Dynamics*, vols 1 and 2 (Cambridge University Press, Cambridge, UK, 1991)
22. M.E. Koepke, W.E. Amatucci, J.J. Carol III, T.E. Sheridan, Experimental verification of the inhomogeneous energy-density driven instability. Phys. Rev. Lett. **72**, 3355–3358 (1994)
23. L.D. Landau, E.M. Lifshitz, *Fluid Mechanics*, 2nd edn. (Pergamon Press, Oxford, UK, 1987)
24. B. Mandelbrot, in *The fractal geometry of nature* (updated and augm. ed.) (W. H. Freeman, New York, USA, 1983)
25. G.V. Munceleanu, V.P. Paun, I. Casian-Botez, M. Agop, The microscopic-macroscopic scale transformation through a chaos scenario in the fractal space-time theory. Int. J. Bif. Chaos **21**, 603–618 (2011)
26. P. Nica, P. Vizureanu, M. Agop, S. Gurlui, C. Focsa, N. Forna, P.D. Ioannou, Z. Borsos, Experimental and theoretical aspects of aluminium expanding laser plasma. Jpn. J. Appl. Phys. **48**, 066001 (2009)
27. P. Nica, M. Agop, S. Gurlui, C. Bejinariu, C. Focsa, Characterization of aluminium laser produced plasma by target current measurements. Jpn. J. Appl. Phys. **51**, 106102 (2012)

28. L. Nottale, *Fractal Space-Time and Microphysics: Towards A Theory Of Scale Relativity* (World Scientific, Singapore, 1993)
29. L. Nottale, *Scale relativity and fractal space-time: a new approach to unifying relativity and quantum mechanics* (Imperial College Press, London, UK, 2011)
30. J.J. Rasmussen, R. Schrittwieser, On the current-driven electrostatic ion-cyclotron instability: a review. IEEE Trans. Plasma Sci. **19**, 457–501 (1991)
31. S. Samukawa, M. Hori, S. Rauf, K. Tachibana, P. Bruggeman, G. Kroesen, J. Christopher Whitehead, A.B. Murphy, A.F. Gutsol, S. Starikovskaia, U. Kortshagen, J.P. Boeuf, T.J. Sommerer, M.J. Kushner, U. Czarnetzki, N. Mason, The 2012 plasma roadmap. J. Phys. D: Appl. Phys. **45**, 253001 (2012)
32. M. Sanduloviciu, Quantum processes as generators of the energy source for ion-cyclotron oscillations. Rev. Roum. Phys. **32**, 745–756 (1987)
33. M. Sanduloviciu, E. Lozneanu, On the generation mechanism and the instability properties of anode double layers. Plasma Phys. Control. Fusion **28**, 585–595 (1986)
34. N. Sato, R. Hatakeyama, A mechanism for potential-driven electrostatic ion-cyclotron oscillations in plasma. J. Phys. Soc. Japan **54**, 1661–1664 (1985)
35. R. Schrittwieser, Modulation of the current-driven ion-cyclotron instability by the potential relaxation instability. Phys. Fluids **26**, 2250–2255 (1983)

Theoretical Modeling of the Interaction Between Two Complex Space Charge Structures in Low-Temperature Plasma

Stefan Irimiciuc, Dan-Gheorghe Dimitriu and Maricel Agop

1 Introduction

Discharge plasmas can be assimilated to complex systems taking into account their structural-functional duality [2, 23]. The standard models [fluid model, kinetic model etc. [7, 24] used to study the discharge plasma dynamics are based on the hypothesis, otherwise unjustified, on the differentiability of the physical variables that describes it, such as energy, momentum, density, etc. The success of the differentiable models must be understood sequentially, i.e. there are domains large enough for the differentiability to be valid.

But differential methods fail when facing the physical reality. Instabilities of the discharge plasmas that can generate chaos or patterns through self-structuring must be analyzed by means of the non-differentiable (fractal) methods [4, 12, 13, 18, 20].

In order to describe some of the dynamics presented in a plasma discharge by means of the non-differentiable method, and still remain treatable as differential method, it is necessary to introduce, the scale resolution, both in the expressions of the physical variables and the dynamics equations. This means that any dynamic variable dependent, in a classical meaning, on the spatial coordinates and time, become dependent also, in a non-differential meaning, on the scale resolution. In other words, instead of working with a dynamic variable, described by means of a mathematical function strictly non-differentiable, we will work just with the different approxi-

S. Irimiciuc · D.-G. Dimitriu (✉)
Faculty of Physics, "Alexandru Ioan Cuza" University, Iasi, Romania
e-mail: dimitriu@uaic.ro

S. Irimiciuc
e-mail: stefan.irimiciuc@yahoo.com

M. Agop
Department of Physics, Faculty of Machine Manufacturing and Industrial Management, "Gheorghe Asachi" Technical University, Iasi, Romania
e-mail: m.agop@yahoo.com

C. H. Skiadas (ed.), *Fractional Dynamics, Anomalous Transport and Plasma Science*, https://doi.org/10.1007/978-3-030-04483-1_6

mation of the variable, derived though its averaging at different scale resolutions. Consequently, any dynamic variable acts as the limit of a family of functions, the functions being non-differentiable for a non-zero scale resolution and differentiable for a null scale resolution [22, 26, 27].

This approach, well adapted to the discharge plasmas dynamics, where any real determination is conducted at a finite scale resolution, clearly implies the development of both a new geometric structure and a physical theory applied to discharge plasmas dynamics, for which the motion laws, invariant to spatial and temporal coordinate transformations, need to be completed with scale laws, invariant to the scale transformations. Such a physical theory that includes the geometric structure based on the above presented assumptions was developed both in the Scale Relativity Theory with fractal dimension 2 [26, 27] and in the extended Scale Relativity Theory, i.e. the Scale Relativity Theory with an arbitrary constant fractal dimension [11, 22]. In the field of discharge plasma, if we assume that the complexity of the interactions in the plasma volume is replaced by non-differentiability (fractality), the constrained motions on continuous but differentiable curves in a Euclidian space of the discharge plasmas particles are replaced with the free motions, without any constrains, on continuous but non-differentiable curves in a fractal space of a "special fluid" (we call it fractal fluid). Thus, the deterministic trajectories are replaced by a collection of potential states, so that the concept of "definite position" is substituted by that of a "definite probability density". As a consequence, the determinism and potentiality (non-determinism) become distinct parts of the same "evolution", through reciprocal interactions and conditioning, in such way that the plasma discharge particles are substituted with the geodesics themselves [8, 14, 26, 27].

2 Theoretical Investigations of the Electronic Oscillations in Discharge Plasmas

Let us assume that a discharge plasma can behave like a multifractal system. Some facets of this statement were previously discussed in [11, 25]. Furthermore, any of the plasma particles, either we discus about the electrons, atoms, or ions, are in permanent interactions in the plasma volume. In-between two collisions, the trajectory of a particle is assumed to be a straight line, that describes the differential framework of the movement, while in the impact point it is a "pre-fractal" that could specify the non-differentiable character of the movement. Considering that all the collisions impact points are forming an uncountable set of points, it results that the trajectories become fractals. So, in the framework of the collective movement of the discharge plasma particles, different classes of trajectories are appearing and thus, different fractals associated to these trajectories simultaneously exist. Therefore, from a global perspective the discharge plasma appears to behave like a multifractal system (for details on the concept of multifractality we recommend the following reference [4, 12, 13, 18, 20].

In such a conjecture, the presence of external constrains on the discharge plasma implies, by means of the fractal theoretical models, analysis of the particle dynamics at particular scale resolutions. In this way we will select both the trajectories classes and the type of fractal characterizing the movement of the particles. Since, generally, the fractality is of stochastic type [20, 26, 27], by applying an external constrains to the discharge plasma, we are also able to select a certain stochastic distribution which is compatible with the ab initio constrains. More precisely, we define the average values of the physical variables with respect to the distribution that defines the class of the measurable physical quantities for a particular experiment.

Let us reconsider the previously mentioned formalism and analyze the dynamics of a discharge plasma on which an external constrains is applied (for example a positive voltage applied to an electrode immersed into plasma).

By means of such constrains, from all the possible dynamics of the discharge plasma particles, only those compatible with these constrains are selected. These dynamics can be "seen" through local Ohm law as variations of the macroscopic conduction current density J [16, 35]:

$$\frac{\mu\varepsilon}{\sigma}\frac{\partial^2\overline{\boldsymbol{J}}}{\partial t^2}+\mu\frac{\partial\overline{\boldsymbol{J}}}{\partial t}=\frac{1}{\sigma}\Delta\overline{\boldsymbol{J}}-\nabla\cdot\left(\nabla\cdot\overline{\boldsymbol{J}}\right), \tag{1}$$

where μ is the magnetic permeability, ε is the electrical permittivity and σ is the electrical conductivity of the discharge plasma, Δ is the Laplace operator, ∇ the gradient operator and ∇ is the divergence operator.

By applying now the variable separation method from [5] in Eq. (1):

$$J^l(x,y,z,t)=J_1^l(x,y,z)J_2(t), \tag{2}$$

we obtain:

$$\frac{1}{\sigma}\left[\partial^i\partial_i J_1^l-\partial^l\partial_i J_1^i\right]+\lambda J_1^l=0, \tag{3}$$

$$\frac{\mu\varepsilon}{\sigma}\frac{\partial^2 J_2}{\partial t^2}+\mu\frac{dJ_2}{dt}+\lambda J_2-0, \tag{4}$$

where $\lambda = \text{const.} > 0$ is a variable separation constant.

By applying Eq. (1) with the explicit forms (Eqs. 3 and 4) to the physical phenomena from our experiment, we need to take into account the following:

(i) Considering the "*restrictions*" imposed by our experiment we will have information only about the temporal evolution of the current density, i.e. $J_2(t)$ is assimilated to the discharge current, while the spatial dependences are disregarded.
(ii) The introduction of external restrictions for the discharge plasma dynamics will be analyzed as follows: the dynamics of an electron beam accelerated by an electrical field which impinges into a neutral medium is analyzed. As a results

of these restrictions, ionizations are produced both by the primary electrons (from the beam) which are accelerated by the electric field $\alpha J_0 = \text{const.}$—with α a coefficient associated with the primary ionization process and J_0 the beam current density—and the secondary electrons which result from the ionization processes, $\beta J_0 \rho_e$—with β a coefficient associated with the secondary ionization process and ρ_e is the electron density. As such, the electric field perceived by the secondary electrons is weaker than that of the primary ones. The contribution of the secondary electrons to the overall discharge plasma dynamics can be negligible so that we can impose $\beta J_0 \rho_e \to 0$.

(iii) Electron density is much higher than that of the ions; we will further focus on the study of the discharge plasma dynamics induced only by the electron branch by means of Eq. (4).

(iv) The multi-fractal character of the discharge plasma dynamics will be introduced by means of structural constant dependencies: $\alpha, \sigma, \mu, \varepsilon$ on the fractality degree $\Lambda = \lambda_0 (dt)^{(2/D_F)-1}$ as follows:

$$\alpha = \alpha(\Lambda), \quad \mu = \mu(\Lambda), \quad \varepsilon = \varepsilon(\Lambda), \quad \sigma = \sigma(\Lambda), \tag{5}$$

where, λ_0 is the coefficient associated to the fractal-non-fractal resolution scale, dt is the resolution scale and D_F is the fractal dimension of the movement curve of a plasma particle. Thus, in the discharge plasma, movements on fractal curves with different associated fractal dimensions (differences induced by their density and kinetic/thermal energy) simultaneously exist, which impose a multifractal character. Let us notice that for $D_F = 2$ the discharge plasma dynamics have a quantum characters, for $D_F < 2$ the dynamics are correlative, while for $D_F > 2$ the dynamics are non-correlative [20, 26, 27].

Therefore, in such context both chaos and self-structuring become fundamental *states* in the evolution of the discharge plasma. Considering the previously made considerations, the dynamics of discharge plasmas can be given by the following evolution equation:

$$\frac{\mu\varepsilon}{\sigma}\frac{\partial^2 J_2}{\partial t^2} + \mu\frac{dJ_2}{dt} + \lambda J_2 = 0, \tag{6}$$

and by using the following substitutions:

$$M = \frac{\mu\varepsilon}{\sigma}, \quad 2R = \mu, \quad K = \lambda, \quad \delta J_3 = \lambda J_2 - \alpha J_0, \quad J_3 = q \tag{7}$$

Equation (6) will take the form of a damped harmonic oscillator:

$$M\ddot{q} + 2R\dot{q} + Kq = 0. \tag{8}$$

Then, M becomes the equivalent of the rest mass, R of the damping coefficient and K of the structure constant. Written in the form:

$$\dot{p} = -\frac{2R}{M}p - \frac{K}{M}q, \tag{9a}$$

$$\dot{q} = p, \tag{9b}$$

Equation (9a) induces a two-dimensional manifold of "phase space" type (p, q) in which p would correspond to a "momentum" type variable and q to a "position" type variable. Equation (9b) corresponds only to a definition of the "momentum". However, Eqs. (9a) and (9b) do not represent a Hamiltonian system, since the associated matrix is not an involution (the matrix trace is not null). The statement becomes more obvious if we put our system in its matrix form:

$$\begin{pmatrix} \dot{p} \\ \dot{q} \end{pmatrix} = \begin{pmatrix} -2\frac{R}{M} & -\frac{K}{M} \\ 1 & 0 \end{pmatrix} \begin{pmatrix} p \\ q \end{pmatrix}. \tag{10}$$

As long as M, R, K have constant values, this matrix equation evidences the position of the energy and thus of the Hamiltonian (obviously only for the cases in which the energy can be identified with the Hamiltonian). Indeed, from Eq. (10) we can obtain:

$$\frac{1}{2}M(p\dot{q} - q\dot{p}) = \frac{1}{2}\left(Mp^2 + 2Rpq + Kq^2\right), \tag{11}$$

which proves that the energy in its quadratic form (the right-hand side of Eq. 11) is the variation rate of the physical action represented by the elementary area from the "phase space". We would like to showcase here the fact that the energy does not have to obey the conservation laws such as the variation rate of the physical action. One can ask now what could be the conservation laws, if they exist, for such a system. To give an adequate answer we first observe that Eq. (11) can be written as a Riccati type differential equation:

$$\begin{aligned} &\dot{\nu} + \nu^2 + 2\bar{\lambda}\nu + \omega_0^2 = 0\ , \\ &\nu = \frac{p}{q},\ \bar{\lambda} = \frac{R}{M},\ \omega_0^2 = \frac{K}{M}\ . \end{aligned} \tag{12}$$

Furthermore, let us note that the Riccati type Eq. (12) always represents a Hamiltonian system describing harmonic oscillator type dynamic:

$$\begin{pmatrix} \dot{p} \\ \dot{q} \end{pmatrix} = \begin{pmatrix} -\frac{R}{M} & -\frac{K}{M} \\ 1 & \frac{R}{M} \end{pmatrix} \begin{pmatrix} p \\ q \end{pmatrix}. \tag{13}$$

This is a general characteristic describing the Riccati type equations and the Hamiltonian's dynamics [1, 19]. Equation (11) can be reobtained by building from matrix Eq. (13) the 1-differential form of the elementary area from the "phase space".

Moreover, the same Eq. (11) can be directly integrated in order to describe the conservation law [10]:

$$\frac{1}{2}\left(Mp^2 + 2Rpq + Kq^2\right)\exp\left\{\frac{2R}{\sqrt{MK - R^2}}\arctan\left(\frac{Mp + Rq}{q\sqrt{MK - R^2}}\right)\right\} = const. \tag{14}$$

From here, it results that the energy is conserved in a classical meaning either if R becomes null, or if the movement in the "phase space" is characterized by the straight line passing though origin, having the slope defined by the ratio between R and M.

Let us now note that Eq. (14) can be rewritten in the form:

$$\frac{Kq^2}{2} = \frac{const}{1 + 2rw + w^2}\exp\left\{2\frac{r}{\sqrt{1 - r^2}}\arctan\left(\frac{w\sqrt{1 - r^2}}{1 - rw}\right)\right\}.$$

$$w^2 = \frac{Mp^2}{Kq^2},\quad r^2 = \frac{R}{kM} \tag{15}$$

This shows in an explicit manner that the potential energy, ignoring a constant factor, depends on the ratio between the kinetic energy and the potential one of a "local" oscillator. Moreover, the similarities between Eq. (15) and the distribution function of a pre-established ensemble of "local oscillators" associated to the thermal radiation [21]:

$$P(r, w) = \frac{1}{1 + 2rw + w^2}\exp\left\{\frac{2r}{\sqrt{1 - r^2}}\arctan\left(\frac{w\sqrt{1 - r^2}}{1 - rw}\right)\right\}, \tag{16}$$

where r is the correlation coefficient and $w^2 = \varepsilon_0 / u$ is the ratio between the thermal energy quanta, ε_0, and the "reference energy", u, can emphasis the statistic character of the energy, namely the ratio of the two types of energies becomes more of a statistic variable. The previously mentioned similarities work under the assumption of quantization type procedure [21]:

$$P(r, w = 1) = \frac{1}{2(1 + r)}\exp\left\{\frac{2r}{\sqrt{1 - r^2}}\arctan\left(\frac{\sqrt{1 - r^2}}{1 - r}\right)\right\}, \tag{17}$$

It can be seen in Fig. 1 that the correlation of the statistic ensemble associated with the local oscillators, previously defined [15, 21]:

$$\exp\left(-\frac{\varepsilon_0}{u}\right) \equiv P(r, w = 1), \tag{18}$$

induces, in the limit of $r \to 0$, the action quanta:

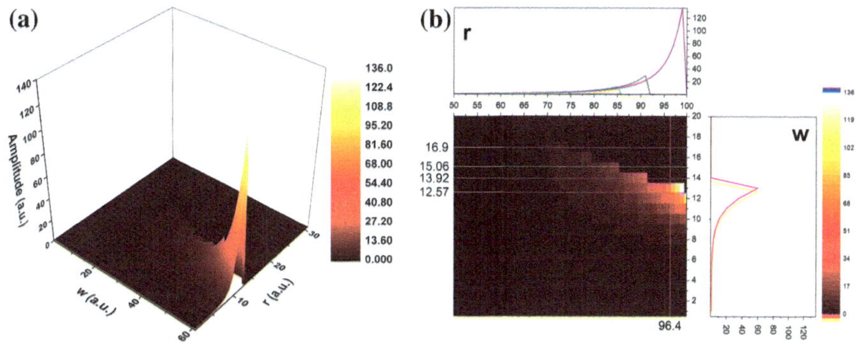

Fig. 1 "Quantization" procedure through "coherence" of all statistical ensembles associated with "local oscillators": dependences 3D (**a**) and contour plot (**b**)

$$\varepsilon_0 = u \ln 2. \tag{19}$$

By expanding of the reference energy, u, in the form:

$$u = k_B T = 2m_0 \lambda_0 (dt)^{(2/D_F)-1} \nu, \tag{20}$$

Equation (19) becomes:

$$\varepsilon_\nu = k_B T \ln 2 = 2m_0 \lambda_0 (dt)^{(2/D_F)-1} \nu \ \ln\ 2. \tag{21}$$

The action quanta and its associated frequency, become proportional with the temperature. In the previous relations k_B is the Boltzmann constant, T is the characteristic temperature of the thermal radiation, ν is the thermal radiation frequency and m_0 is the inertial mass of the "local oscillator". Therefore, we can build now, either fractal unitary physical theories, in the de Broglie sense [9], by identifying the fractal physical action with the fractal entropy (like the thermodynamic of isolated particle), or fractal theories based on the fractal information entropy in the Landauer sense [17], regarding the efficiency as a quantum computer, imposed by the fractal physical laws on the fractal computers [3] (particular on the quantum computers). From this last statement, according to the second law of thermodynamics [20], from a fractal perspective the fractal erasure of one fractal bit of fractal information [3] requires, at a given scale resolution, a minimal fractal heat generation of $k_B T \ln 2$, with T being the temperature at which the bit of fractal information can be erased. Since the erasure is a logical function that does not have an inverse fractal single-value, it must be associated with the fractal physical irreversibility and therefore requires fractal heat dissipation. A fractal bit has one degree of freedom and so the heat dissipation should be of the order of $k_B T$. Now, since before the erasure a fractal bit can be in any of the two possible states and after the fractal erasure it can only be in one fractal state, this implies a change in fractal information entropy of an amount $k_B ln2$.

Let us now admit that in Eqs. (20) and (21) the fractalization is of stochastics type and it is given by means of Markovian processes [20, 27], i.e. through the movement of the plasma particles on Peano type curves in fractal dimension $D_F = 2$. Then, at Compton scale resolution $\lambda = \hbar/2m_0$, where $\hbar$ is the Planck reduced constant and m_0 is the rest mass of the "local oscillator", Eq. (20) can be reduced to de Broglie equation [9]:

$$u = k_B T = h\nu. \tag{22}$$

Equation (21) leads to the Landauer equation [17]:

$$\varepsilon_\nu = k_B T \ln 2 = h\nu \ln 2. \tag{23}$$

The equation describing the movement of the harmonic oscillation without damping is a stationary property consequence of the average value of the difference between the kinetic energy and the potential one on the whole-time of the movement. Since this difference define the Lagrangian of the movement, its integral in between two moments of time will now define the action. Thus, the action will be proportional with the average value of the difference between the two types of energies. Therefore, the movement of the damped harmonic oscillator distributes the two types of energies in such manner that their mean is stationary during the entire time of the movement.

Not the same considerations as the ones considered before can be applied in order to obtain the Eq. (8). However, they can show the meaning of the statistical argues in order to establish the type of fractalization by stochasticization. For Eq. (8) not the temporal mean of the difference of the two types of energies is stationary, but the temporal mean of the function:

$$L(q, \dot{q}, t) = \frac{1}{2}\left(m\dot{q}^2 - kq^2\right)\exp\left(\frac{2R}{M}t\right). \tag{24}$$

In this case the action of the finite temporal interval given by the integral:

$$S(t_i, t_f) = \int_{t_f}^{t_i} \left(M\dot{q}^2 - kq^2\right)\exp\left(\frac{2R}{M}t\right)dt, \tag{25}$$

can be again interpreted as the temporal mean of the difference of the two types of energies, but for the a priori statistics of the temporal domain given by an exponential distribution. In the case of the "harmonic oscillator" this a priori distribution is simply uniform. Therefore, one can state that by means of the Hamiltonian of the motion, the equation of the harmonic oscillator is obtained as a consequence of the stationarity of a temporal statistic. The difference between them consists only in the a priori measure of the temporal axis specific to the analyzed dynamics.

Let us observe that the motion Eq. (8) cannot be directly obtained through the variational principle applied to the action (25), $\delta S(t_i, t_f) \equiv 0$. Supplementary constraints are necessary at the limits:

(i) the variation of the coordinate at the time interval limits is null:

$$\delta q|_{t_i} = \delta q|_{t_f} \equiv 0, \tag{26}$$

(ii) for the closed trajectories the limit values of the coordinates at the interval limits must be the same:

$$q(t_i) = q(t_f), \tag{27}$$

(iii) for the closed trajectories, but in the "phase space", the values of the velocities corresponding to the initial and final states should be the same:

$$\delta \dot{q}(t_i) = \delta \dot{q}(t_f). \tag{28}$$

Even if we consider them, one can verify from the variational principle perspective, $\delta S(t_i, t_f) \equiv 0$, that the Lagrangian is not unique. In other words, the Lagrangian is defined up to an arbitrary function (the time-derivative of a function) with equal values of the time limits intervals. In such context, it is necessary a gauge procedure (of the Lagrangian uniqueness) by means of which the Lagrangian is reduced to a perfect square. The Riccati type equation becomes then only the necessary gauge condition, it also specifying the way in which the statistics of the harmonic oscillator must be described.

The procedure [36] consists in the adding the following term to the Lagrangian (24)

$$\frac{1}{2}\frac{d}{dt}\left[wq^2 \exp\left(2\frac{R}{M}t\right)\right], \tag{29}$$

where w is a continuous function in time, imposing that the Lagrangian should be a perfect square. The function variation given by the derivative operator is null, due to the conditions presented in Eq. (27), thus the motion equation does not change. The new Lagrangian, written in relevant coordinates, takes the form:

$$L(\dot{q}, q, t) = \frac{M}{2}\left(\dot{q} + \frac{w}{M}q\right)^2 \exp\left(\frac{2R}{M}t\right), \tag{30}$$

with the condition that w needs to satisfy the following Riccati type equation:

$$\dot{w} - \frac{1}{M}w^2 + 2\frac{R}{M}w - K = 0, \tag{31}$$

The Lagrangian depicted in Eq. (30) will be considered here as representing the energy of the analyzed system. As before, there is a relationship between the Riccati type Eq. (31) and the Hamiltonian dynamics. Henceforth we will find a similar relation to that one presented in Eq. (13):

$$\begin{pmatrix} \dot{\eta} \\ \dot{\xi} \end{pmatrix} = \begin{pmatrix} -\frac{R}{M} & -\frac{K}{M} \\ 1 & \frac{R}{M} \end{pmatrix} \begin{pmatrix} \eta \\ \xi \end{pmatrix}, \quad w = \frac{\eta}{\xi}, \tag{32}$$

This system is obviously a Hamiltonian one. Thus, we can identify the factors of w with the "phase space" coordinates. It is important to find the most general solution of the Eq. (27). José Cariñena and Arturo Ramos presented a modern approach to integrate a Riccati type equation [6]. For our case is enough to note that the complex numbers:

$$w_0 \equiv R + iM\Omega, \quad w_0^* \equiv R - iM\Omega, \tag{33}$$

with

$$\Omega^2 = \frac{K}{M} - \left(\frac{R}{M}\right)^2, \tag{34}$$

are the roots of the quadratic polynomial $P(\nu) = \frac{1}{M}\nu^2 - 2\frac{R}{M}\nu + K = 0$. Thus, making first the homographic transformation:

$$z = \frac{w - w_0}{w - w_0^*}, \tag{35}$$

it results by direct determination that z is a solution of the linear and homogeneous first order equation:

$$\dot{z} = 2i\Omega z \therefore z(t) = z(0)e^{2i\Omega t}. \tag{36}$$

Hence, if we express the initial condition $z(0)$ in a right manner, we can obtain the general solution of Eq. (31) by applying an inverse transformation to Eq. (34). We find:

$$w = \frac{w_0 + w_0^* r e^{2i\Omega(t-t_r)}}{1 + r e^{2i\Omega(t-t_r)}}, \tag{37}$$

where r and t_r are two real constants which characterize the solution. Using Eq. (33), we can put the same solution in real terms in the form:

$$z = R + M\Omega\left(\frac{2r\sin[2\Omega(t-t_r)]}{1 + r^2 + 2r\cos[2\Omega(t-t_r)]} + i\frac{1 - r^2}{1 + r^2 + 2r\cos[2\Omega(t-t_r)]}\right). \tag{38}$$

This relationship shows a frequency modulation through a Stoler transformation [33] which leads to the complex representation of this parameter.

Let us note that in the quantum field theory the Stoler transformation it is related to the creation/annihilation from the second quantization. In this work these transformation specify the fact that the collisions processes which can be found in our discharge plasma implicitly are seen as *creation* and *annihilation* of charged particles. Moreover the correlation with *friction* behaves as a frequency self-modulation phenomena. This process acts as a calibration between the kinetic energy and the potential one, as is usually the Lagrangian defined, which reduced this parameter to a *perfect square*. The physical significance of the squared Lagrangian is that of a fundamental physical unit that describes the inner structures of a complex physical system, i.e. in our case the two structures that are both conditioning each other.

Through Eq. (31) w becomes a variation rate of the electrical charge which allows us now to evaluate, in an indirect manner, the variation in the discharge current. In the following we will present the evolution of the discharge current at various resolution scales, in term of frequency, as function of time and oscillation frequency that modulates the oscillations. The results are presented in Fig. 2.

The theoretical model proposed here attempts to explain the dynamics of charged particles in a plasma discharge where there is a strong flux of electrons from one plasma structure to another. Basically, the dynamics of the electrons is described using a forced damped oscillating system, with the aim to investigate the response of the global discharge current to different changes in resolution scale, oscillation frequency and damping coefficient. Since our mathematical approach is sensitive to the changes in the resolution scales, we plotted in Fig. 2 the 3D maps and the corresponding contour plot representations of the discharge current as functions of time and oscillation frequency for a fixed value of the damping constant of 0.1. We observe that for small resolutions the current is described by a simple oscillatory regime, while as the frequency resolution scale increases we notice the appearance of some patterns. The patterns become denser and are foreshadowing the presence of some modulation of the oscillating frequency.

The damping of the oscillatory state describes the losses through dissipative or dispersive mechanisms. In order to study the effects induced by these mechanisms on the global current, we pin pointed two different oscillation frequency (characterizing one the plasma structure), and observe the temporal response to different values of the damping coefficient. The results can be seen in Fig. 3. We can identify competing oscillatory behaviors described by two oscillation frequencies with comparable amplitudes.

The effect of the forced oscillations that can be attributed to one of the plasma structures is presented in Fig. 4 where the discharge current for a fixed value of the damping coefficient and various values of the forced oscillations is shown. The system seemingly starts from a state described by period doubling, and goes through frequency modulation as the values increases. The important aspect is that the oscillation frequencies found for the current are not the ones induced through forced oscillatory system. This means that the system, although forced to get on a specific

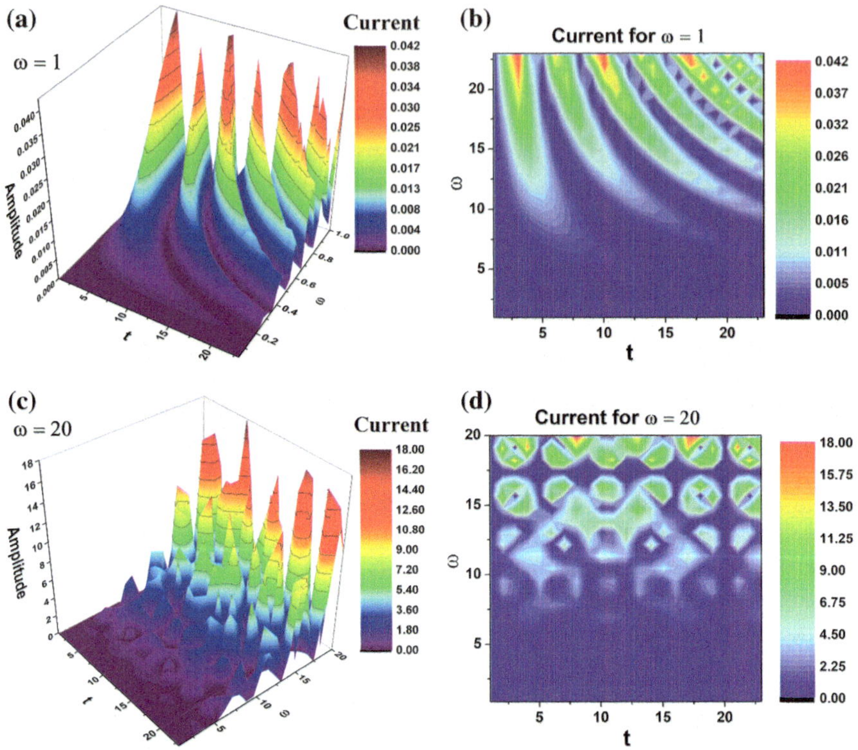

Fig. 2 3D representation of the discharge current for different resolution scales in respect to the oscillation frequency (**a** ~0–1, **c** ~0–20) and the corresponding contour plot representations (**b** and **d**, respectively)

state, it will define its own dynamics influenced but not determined by the external parameters.

3 Experimental Investigations of Space Charge Structures Generated in a Spherical Cathode with an Orifice

The results obtained from the theoretical model described above were experimentally confirmed by the results related to the nonlinear dynamics of a very complex space-charge configuration appearing inside and around a spherical grid cathode with hole, extensively described in [28, 34]. The experiments have been performed in a plasma diode made by glass, where the anode is a rectangular plate of 25 cm × 20 cm size, while the cathode is a spherical metallic grid with a diameter of 4 cm (with 0.5 mm being the diameter of the metallic wires and 2 mm the mesh width) and with a small hole with a diameter of about 6 mm on one side. The distance between cathode and

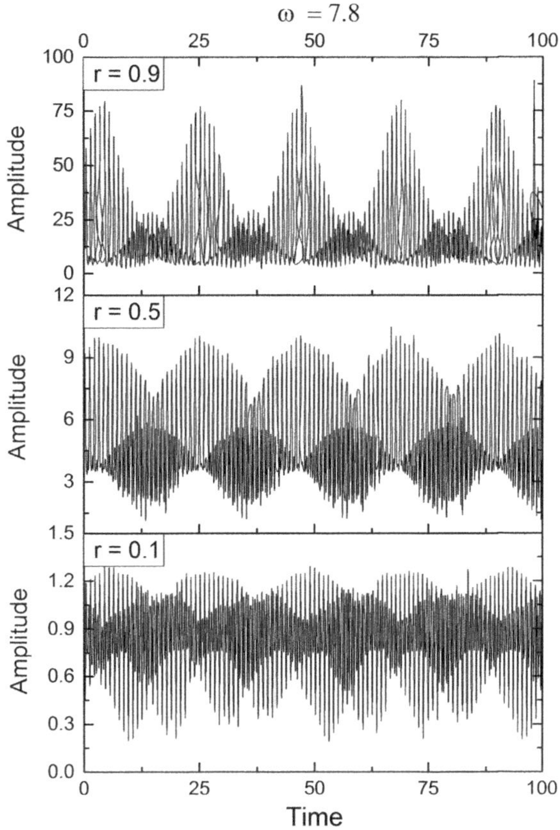

Fig. 3 Discharge current temporal traces for various values of the damping coefficient

anode is d = 25 cm. For experiments, Argon has been used as working gas, at the pressure p = 7 × 10–2 mbar.

By modifying the discharge voltage, the complex space-charge configuration inside and around the cathode passes through different stages. Thus, at low values of the discharge voltage, a diffuse spherical structure appears inside the cathode, while a strong electrons beam escape from inside the cathode through the hole, producing excitations and ionizations of the gas atoms along the direction of propagation (see photo in Fig. 5a). In this stage, there is no breakdown of the discharge, but the increasing of the current is due only to these electrons escaping from inside the cathode through the hole, which produce local ionizations of the gas atoms along the direction of propagation. The plasma bubble inside the cathode [29–32] appears due to the electron-neutral impact excitations and ionizations processes, a high density of electrons existing there because of the hollow cathode effect.

For high values of the discharge voltage, after the breakdown of the discharge, the negative glow of the discharge can be observed in Fig. 5b, which is coupled with the bubble inside the cathode through a fireball in form of an asymmetric dumbbell (let's

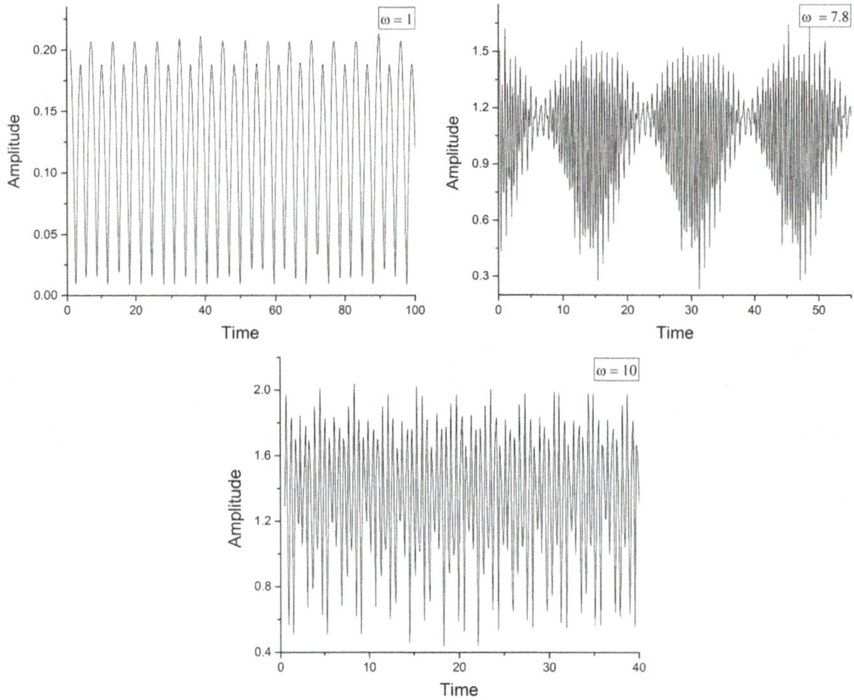

Fig. 4 Discharge current temporal traces for different values of the oscillation frequency at a constant damping coefficient ($r = 0.1$)

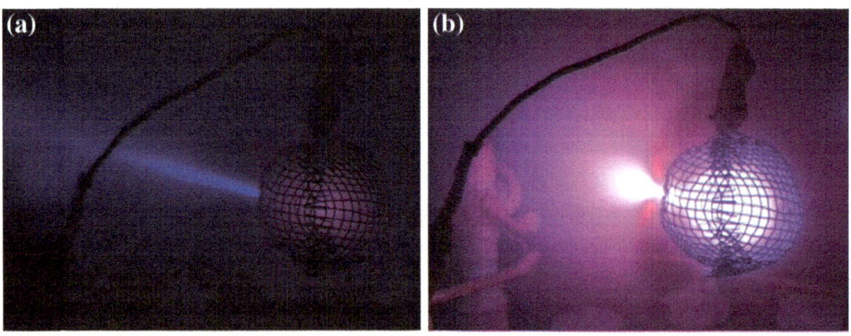

Fig. 5 Photos of the complex space-charge configuration around and inside the cathode in different stages of its development

call this as firedumbbell). The right part (towards the cathode) of this firedumbbell is smaller and penetrates inside the cathode through the hole, while its left part (towards the negative glow of the discharge) is larger and diffuse into the negative glow. In this state interesting nonlinear dynamic phenomena were observed by investigating the time series of the discharge current oscillations.

Time series of the discharge current oscillations were recorded with a sampling rate of 2.5 GS/s. Figure 6 shows such a time series, recorded for a discharge voltage value $U = 289$ V. Figure 7 shows details from this time series of the discharge current oscillations. Different dynamic states can be observed, many of them being similar

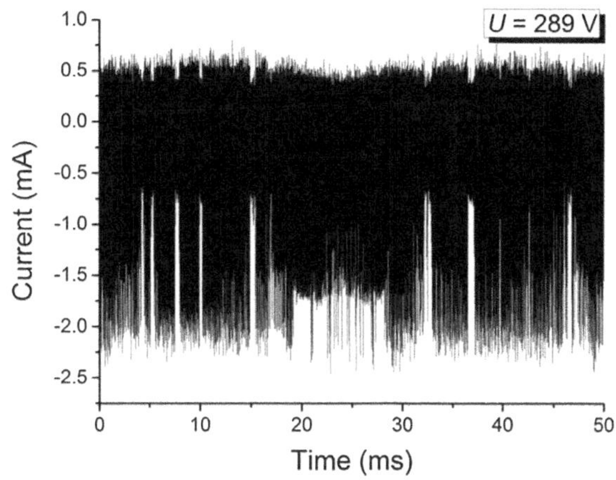

Fig. 6 Time series of the discharge current oscillations for the value of the discharge voltage $U =$ 289 V

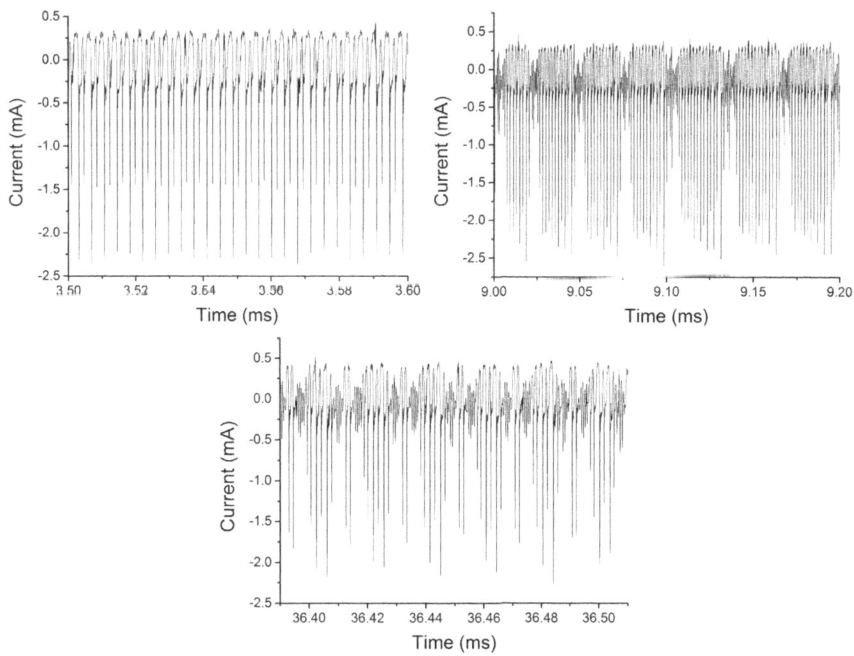

Fig. 7 Details of the discharge current oscillations from Fig. 6

with those provided by the theoretical model described in the previous section. The discharge current oscillations are the results of the nonlinear interaction between the inner bubble and the firedumbbell. Both structures are in dynamic states, but the oscillation frequencies depend on the current [25, 29–32]. The inner bubble periodically release bunches of electrons, which act as forcing drive of the oscillating firedumbbell. As the theoretical model predicted, the firedumbbell passes through different dynamic state as the forcing frequency change because of the modification of the discharge current value. In other words, the forcing frequency determines the dynamic state, i.e. the discharge current oscillations, which, at their turn, modify the frequency of the inner bubble dynamics, i.e. the forcing frequency. In this way, a continuous self-modulation of the discharge current arises, as can be observed in Fig. 6 and in details in Fig. 7.

4 Conclusion

Based on a non-differential approach, an explanation was proposed for the modulated oscillation of plasma structured created in a spherical cathode with an orifice. Within the framework of the model the particles move on fractal curves which might lead to an oscillating state of the charged particles. The system evolves from a double period state towards a chaotic signal but never reaching it. The evolution of the system is "controlled" by the damping of the system and by the maximum frequency reached during a particular simulated time-series. The experimental data depicting the discharge plasma current for a constant positive potential applied on the spherical cathode reveals an oscillating behavior. For a time-series, the oscillations are modulated with various frequencies, results also captured by the theoretical simulations.

Acknowledgements This work was supported by a grant of Romanian Ministry of Research and Innovation, CNCS—UEFISCDI, project number PN-III-P4-ID-PCE-2016-0355, within PNCDI III.

References

1. V.I. Arnold, *Mathematical Methods of Classical Mechanics*, 2nd edn. (Springer, New York, USA, 1989)
2. R. Badii, A. Politi, *Complexity: Hierarchical Structure and Scaling in Physics* (Cambridge University Press, Cambridge, UK, 1997)
3. V. Barlescu, M. Agop, M. Craus, Computational properties of a fractal medium. Int. J. Quantum Inf. **12**, 1450022 (2014)
4. M.F. Barnsley, *Fractals Everywhere*, 2nd edn. (Academic Press, Cambridge, MA, USA, 1993)
5. A.V. Bitsazde, *Equations of Mathematical Physics* (Mir Publishers, Moscow, Russia, 1980)
6. J.F. Cariñena, A. Ramos, Integrability of Riccati equations from a group-theoretical viewpoint. Int. J. Mod. Phys. A **14**, 1935–1951 (1999)
7. F.F. Chen, *Introduction to Plasma Physics and Controlled Fusion*, 3rd edn. (Springer International Publishing, Cham, Switzerland, 2016)

8. C.P. Cristescu, *Nonlinear dynamics and chaos. Theoretical Fundaments and Applications* (Romanian Academy Publishing House, Bucharest, Romania (in Romanian), 2008)
9. L. de Broglie, *La Thermodynamique de la Particule Isolée* (Gauthier-Villars, Paris, France, 1964)
10. H.H. Denman, Time-translation invariance for certain dissipative classical systems. Am. J. Phys. **36**, 516–519 (1968)
11. D.G. Dimitriu, S.A. Irimiciuc, S. Popescu, M. Agop, C. Ionita, R.W. Schrittwieser, On the interaction between two fireballs in low temperature plasma. Phys. Plasmas **22**, 113511 (2015)
12. K. Falconer, *Fractal Geometry: Mathematical Foundations and Applications* (John Wiley & Sons, Chichester, UK, 2014)
13. H.M. Hastings, G. Sugihara, *Fractals: A User's Guide for the Natural Sciences* (Oxford University Press, Oxford, UK, 1994)
14. R.C. Hilborn, *Chaos and Nonlinear Dynamics: An introduction for Scientists and Engineers*, 2nd edn. (Oxford University Press, Oxford, UK, 2000)
15. H. Ioannidou, Statistical interpretation of the Planck constant and the uncertainty relation. Int. J. Theor. Phys. **22**, 1129–1139 (1983)
16. J.D. Jackson, *Classical Electrodynamics*, 3rd edn. (Wiley, New York, USA, 1999)
17. R. Landauer, Irreversibility and heat generation in the computing process. IBM J. Res. Dev. **5**, 183–191 (1961)
18. A. le Méhauté, *Fractal Geometry: Theory and Applications* (CRC Press, Boca Raton, Florida, USA, 1991)
19. P. Libermann, C.M. Marle, *Symplectic Geometry and Analytical Mechanics* (D. Reidel Publishing Company, Dordrecht, Holland, 1987)
20. B.B. Mandelbrot, *The Fractal Geometry of Nature* (W.H. Freeman and Company, New York, USA, 1982)
21. N. Mazilu, M. Agop, *At the Crossroads of Theories* (Between Newton and Eistein – Barbilian Universe, Ars Longa, Iasi, Romania (in Romanian), 2010)
22. I. Merches, M. Agop, *Differentiability and Fractality in Dynamics of Physical Systems* (World Scientific, Singapore, 2016)
23. M. Mitchell, *Complexity: A Guided Tour* (Oxford University Press, Oxford, UK, 2009)
24. A.I. Morozov, *Introduction to Plasma Dynamics* (CRC Press, Boca Raton, Florida, USA, 2012)
25. O. Niculescu, D.G. Dimitriu, V.P. Paun, P.D. Matasaru, D. Scurtu, M. Agop, Experimental and theoretical investigations of a plasma fireball dynamics. Phys. Plasmas **17**, 042305 (2010)
26. L. Nottale, *Fractal Space-Time and Microphysics: Towards a Theory of Scale Relativity* (World Scientific, Singapore, 1993)
27. L. Nottale, *Scale Relativity and Fractal Time-Space: A New Approach to Unifying Relativity and Quantum Mechanics* (Imperial College Press, London, UK, 2011)
28. R.W. Schrittwieser, C. Ionita, C.T. Teodorescu-Soare, O. Vasilovici, S. Gurlui, S. Irimiciuc, D.G. Dimitriu, Spectral and electrical diagnosis of complex space-charge structures excited by a spherical grid cathode with orifice. Phys. Scr. **92**, 044401 (2017)
29. R.L. Stenzel, J.M. Urrutia, Oscillating plasma bubble. I. Basic properties and instabilities. Phys. Plasmas **19**, 082105 (2012a)
30. R.L. Stenzel, J.M. Urrutia, Oscillating plasma bubble. II. Pulsed experiments. Phys. Plasmas **19**, 082106 (2012b)
31. R.L. Stenzel, J.M. Urrutia, Oscillating plasma bubble. III. Internal electron sources and sinks. Phys. Plasmas **19**, 082107 (2012c)
32. R.L. Stenzel, J.M. Urrutia, Oscillating plasma bubble. IV. Grids, geometry, and gradients. Phys Plasmas **19**, 082108 (2012d)
33. D. Stoler, Equivalence classes of minimum uncertainty packets. Phys. Rev. D **1**, 3217–3219 (1970)

34. C.T. Teodorescu-Soare, D.G. Dimitriu, C. Ionita, R.W. Schrittwieser, Experimental investigations of the nonlinear dynamics of a complex space-charge configuration inside and around a grid cathode with hole. Phys. Scr. **91**, 034002 (2016)
35. A. Zangwill, *Modern Electrodynamics* (Cambridge University Press, Cambridge, UK, 2013)
36. M.I. Zelikin, in *Control Theory and Optimization I: Homogeneous Spaces and the Riccati Equation in the Calculus of Variations*. Encyclopedia of Mathematical Sciences, vol. 86 (Springer, Berlin, Germany, 2000)

Some Applications of Fractional Derivatives in Many-Particle Disordered Large Systems

Z. Z. Alisultanov, A. M. Agalarov, A. A. Potapov and G. B. Ragimkhanov

1 Introduction

Systems with Hamiltonians containing a nonquadratic degree of momentum have always attracted the attention of researchers. This is due, first of all, to the fact that in real systems, as a result of their imperfection, quadratic dependences of energy on the pulse are almost never realized. On the other hand, a system with a complex dispersion law can not be accurately described within the framework of the existing theory. Therefore, in many cases resort to approximations, which consist in the fact that the nonideal system of particles is represented as a gas of free quasiparticles with a non-quadratic dependence of the energy on the momentum.

The nonquadratic dependence of energy on the pulse leads to quite interesting effects. We indicate here some. It was also noted by Lifshits and Kaganov [1] that in the case of a nonquadratic dependence of the electron energy on the pulse, the

Z. Z. Alisultanov (✉) · A. M. Agalarov
Department of Theoretical Physics, Amirkhanov Institute of Physics of Dagestan Scientific Center of Russian Academy of Sciences, Makhachkala 367015, Russia
e-mail: zaur0102@gmail.com

A. M. Agalarov
e-mail: aglarow@mail.ru

A. A. Potapov
Kotel'nikov Institute of Radio Engineering and Electronics of Russian Academy of Sciences, Moscow 125009, Russia
e-mail: potapov@cplire.ru

A. A. Potapov
JNU-IRAE RAS Joint Laboratory of Information Technology and Fractal Processing of Signals, JiNan University, Guangzhou 510632, China

G. B. Ragimkhanov
Faculty of Physics, Dagestan State University, 367000 Makhachkala, Dagestan Republic, Russia
e-mail: gb-r@mail.ru

C. H. Skiadas (ed.), *Fractional Dynamics, Anomalous Transport and Plasma Science*, https://doi.org/10.1007/978-3-030-04483-1_7

period of revolution of electrons in the magnetic field, and consequently the resonant frequency, depend on the applied electric field. This effect, as the authors of this work pointed out, can be used to analyze the shape of isoenergetic surfaces in semiconductors. Interesting results were obtained in [2], where high-frequency electromagnetic phenomena in semiconductors with a nonquadratic dispersion law of carriers were considered. In [3], electronic chaos in a one-dimensional superlattice was considered. Chaoticity was due to the nonquadratic dispersion law. In [4], nonlinear electromagnetic waves were investigated in a stochastized electron gas with a non-quadratic dependence of the energy on the momentum, and it was shown that a deviation from the quadratic dependence leads to a slowing of the waves.

Another factor that strengthens interest in systems with a non-quadratic dependence of energy on momentum is associated with the penetration into physics of ideas of the so-called fractional calculus based on the mathematical apparatus of integro-differentiation of fractional order [5, 6]. For systems described by differential equations of fractional order, the deviation from the quadratic dependence of the energy on the momentum and the presence of nonlocality of the fractional-power character are characteristic. Such a deviation leads, in particular, to probability distributions of fractional-power type. These distributions turn out to be unusually stable [7, 8]. The use of this apparatus makes it possible to interpret with great accuracy complex experimental data for such phenomena as anomalous diffusion [6], heat transfer in media with a complex structure [9], dispersion transport in semiconductors [10], calculation of thermodynamic properties of surfaces [11]. In addition, using this device, a number of new effects were predicted. For example, it was shown in Refs. [12, 13] that it is possible to achieve comparatively large transition temperatures into the superconducting state within the concept of fractons, which makes interesting the apparatus of fractional integro-differentiation in the theory of superconductivity. It should be noted that non-local interactions of power type have been considered in Dyson's papers as applied to phase transitions in ferromagnets [14–16]. In [17], nonlocal interactions of this type in a crystal lattice were considered and fractional kinetic equations were obtained. The features of fractional quantum mechanics were considered in [18–21], where a study was made of systems with Hamiltonians containing a non-integer degree of momentum. It should be noted that Tarasov made a significant contribution to the generalization of equations to the case of fractal media, as well as the investigation of systems described by equations in fractional operators. For example, in [22–24] fractional statistical mechanics was considered, fractional analogues of the Liouville and Bogolyubov equations were obtained, as well as kinetic equations with fractional derivatives. Fractal analysis has proved extremely useful in radiophysics [25] and in the analysis of complex images [26, 27].

Here we consider some applications of fractional calculus. First, we consider the possibility of introducing fractional time derivatives in the quantum theory of the electromagnetic field. Further, we generalize the equation for the quantum Green's functions to the case of fractional derivatives and investigate some quantum-statistical properties of the many-particle system [28, 29]. Finally, we will consider some specific problems of plasma physics using an approach based on developed formalism [30].

2 The Liouville Fractional Derivative with Respect to Time in Quantum Equations

Here an attempt is made to introduce fractional differentiation with respect to time into the quantum theory of the electromagnetic field and fractional differentiation with respect to the coordinate into the quantum theory of many particles. The starting point of the research is the fractional Maxwell equations, which were introduced into classical electrodynamics [31].

As fractional derivatives, we use the Riemann-Liouville and Caputo integro-differential operators, which are defined as follows [32]

$$\partial_{0\tau}^{\alpha} f(\tau) = sign^{n}(\tau - s) D_{s\tau}^{\alpha-n} \frac{\partial^{n} f(\tau)}{\partial \tau^{n}},$$

where is the Riemann-Liouville operator, defined as

$$D_{s\tau}^{\alpha} f(\tau) = \left\{ \begin{array}{c} \frac{sign(\tau-s)}{\Gamma(-\alpha)} \int\limits_{s}^{\tau} \frac{f(\tau')d\tau'}{|\tau-\tau'|^{\alpha+1}} \\ f(\tau) \\ sign^{n}(\tau - s) \frac{\partial^{n}}{\partial \tau^{n}} D_{s\tau}^{\alpha-n} f(\tau) \end{array} \right\}.$$

2.1 *Representation of the Electromagnetic Field in the Form of a Set of Fractal Oscillators*

Between the electromagnetic field and classical mechanics there is a deep analogy that underlies the quantization of a given field [33]. It is this analogy that shows that an electromagnetic field can be represented as an infinite number of harmonic oscillators. It is of interest to carry out an appropriate investigation for the electromagnetic field described by the fractional Maxwell equations.

It was shown in paper [31] that Maxwell's fractional equations describe the electromagnetic field in media possessing the "memory" property, which consists in the fact that the electromagnetic field in such media is dissipated according to a power law. These results are based on completely realistic considerations and experimental data. Such environments are conventionally called "fractal in time" in many works. We will demonstrate an attempt to quantize the electromagnetic field in such "fractal" media.

In [31] the following fractional Maxwell equations were obtained,

$$\mathrm{rot}\mathbf{E} = -\frac{\mu}{t_0}\partial^{\alpha}_{0\tau}\mathbf{H},$$
$$\mathrm{rot}\mathbf{H} = \frac{\varepsilon}{t_0}\partial^{\alpha}_{0\tau}\mathbf{E}, \tag{2.1}$$

Here, the dimensionless variable is related to the dimensional variable as, where is some characteristic time of the process.

We shall consider a field of relatively low frequencies, at which variance can still be neglected. Therefore, the dielectric and magnetic permeabilities in this case are constants independent of time. The dependence of permeabilities on fields is also neglected, i.e. the fields under consideration are sufficiently small.

The electric and magnetic fields can be represented in the form of an expansion in standing waves. Following the usual theory of secondary quantization [33], we represent the electric and magnetic fields in the form of sums

$$\mathbf{E} = \sum_{\gamma} \omega_\gamma q_\gamma(t)\mathbf{e}_\gamma(r), \quad \mathbf{H} = \sum_{\gamma} p_\gamma(t)\mathbf{h}_\gamma(r). \tag{2.2}$$

The index denotes both indices and, i.e. It includes the polarization of the wave and the direction of the wave vector and, by definition, satisfy the orthonormalization condition [33]

$$\int \mathbf{e}_{\gamma'}(r)\mathbf{e}_\gamma(r)d^3r = \delta_{\gamma'\gamma},$$
$$\int \mathbf{h}_{\gamma'}(r)\mathbf{h}_\gamma(r)d^3r = \delta_{\gamma'\gamma},$$
$$\int \mathbf{e}_{\gamma'}(r)\mathbf{h}_\gamma(r)d^3r = 0. \tag{2.3a}$$

Также справедливо соотношение [11, с. 14]) Equality [11, p. 14]

$$\mathbf{h}_\gamma(r) = \frac{1}{\omega}\mathrm{rot}\mathbf{e}_\gamma(r) \tag{2.3b}$$

Substituting the expressions (2.3b) into Eqs. (2.3a), we obtain the following equations,

$$\sum_{\gamma} \omega_\gamma q_\gamma \mathrm{rot}\mathbf{e}_\gamma(r) = -\frac{\mu}{t_0}\sum_{\gamma} \partial^{\alpha}_{0\tau} p_\gamma(\tau)\mathbf{h}_\gamma(r),$$
$$\sum_{\gamma} p_\gamma \mathrm{rot}\mathbf{h}_\gamma(r) = \frac{\varepsilon}{t_0}\sum_{\gamma} \omega_\gamma \partial^{\alpha}_{0\tau} q_\gamma(\tau)\mathbf{e}_\gamma(r),$$

which, taking into account the relations (2.3a, 2.3b) between u, lead to the following equations of motion

$$\partial_{0\tau}^{\alpha} p_{\gamma} = -\frac{t_0}{\mu}\omega_{\gamma}^2 q_{\gamma},$$
$$\partial_{0\tau}^{\alpha} q_{\gamma} = \frac{t_0}{\varepsilon} p_{\gamma}. \tag{2.4}$$

Because values μ, ε and t_0 in this case do not play a special role, then we write further expressions in the system of units in which $\mu = \varepsilon = t_0 = 1$. For the coordinate, then, we have the equation of motion

$$\partial_{0\tau}^{\alpha}\partial_{0\tau}^{\alpha} q_{\gamma} + \omega_{\gamma}^2 q_{\gamma} = 0.$$

This is the equation of motion for the "fractal" oscillator, studied in detail in [34, 35]. This means that the electromagnetic field described by the fractional Maxwell equations can be represented as a collection of an infinite number of fractal oscillators. Because the fractional order of the derivative in the equation of motion of the fractal oscillator is due to some attenuation of the oscillations inherent in dissipative media, then the fractional order of the time derivative in Maxwell's equations can be interpreted as the presence of dissipation of the electromagnetic field in the medium.

2.2 The Heisenberg Equation

In quantum mechanics, the change in the operator of dynamic quantities with time is described by the Heisenberg equation

$$-i\frac{\partial \hat{F}}{\partial t} = \left[\mathrm{H}, \hat{F}\right], \tag{2.5}$$

where H is the Hamiltonian of the quantum system under consideration.

When quantizing an electromagnetic field, it is shown that Maxwell's equations for the operators of electric and magnetic field strengths are something other than the Heisenberg equations for the corresponding operators. For example,

$$-i\frac{\partial \hat{\mathbf{E}}}{\partial t} = \left[\mathrm{H}, \hat{\mathbf{E}}\right],$$
$$-i\frac{\partial \hat{\mathbf{H}}}{\partial t} = \left[\mathrm{H}, \hat{\mathbf{H}}\right],$$

where H is the Hamiltonian of the electromagnetic field.

We establish the Heisenberg equation for the electromagnetic field described by the fractional Maxwell equations.

Thus, for the electric and magnetic field strength operators, we have the fractional Maxwell Eqs. (3.1). Secondarily quantized electric and magnetic fields are expressed in terms of the operators of generalized coordinates and momenta as follows)

$$\hat{\mathbf{E}} = \sqrt{\frac{2}{V}} \sum_{\nu k} \omega \varepsilon_{\nu \mathbf{k}} \left\{ \hat{q}^{+}_{\nu \mathbf{k}} \cos \mathbf{kr} + \hat{q}^{-}_{\nu \mathbf{k}} \sin \mathbf{kr} \right\},$$

$$\hat{\mathbf{H}} = \sqrt{\frac{2}{V}} \sum_{\nu k} [\mathbf{n}\varepsilon_{\nu \mathbf{k}}] \left\{ -\hat{p}^{+}_{\nu \mathbf{k}} \sin \mathbf{kr} + \hat{p}^{-}_{\nu \mathbf{k}} \cos \mathbf{kr} \right\}. \tag{2.6}$$

where $\varepsilon_{\nu \mathbf{k}}$ is a unit vector $\mathbf{k}$ describing the polarization state of a plane wave with a wave vector. For example, for a circularly polarized wave $\varepsilon_{\nu \mathbf{k}} = \left(1/\sqrt{2}\right)\left(1 \ \nu i \ 0\right)$, $\nu = \pm 1$, where. The Hamiltonian of the electromagnetic field has the following form

$$\mathrm{H} = \frac{1}{2} \int \left(\hat{E}^2 + \hat{H}^2 \right) d^3\mathbf{r} = \frac{1}{2} \sum_{\nu \mathbf{k}} \left(\hat{p}^{+2}_{\nu \mathbf{k}} + \hat{p}^{-2}_{\nu \mathbf{k}} + \left(\hat{q}^{+2}_{\nu \mathbf{k}} + \hat{q}^{-2}_{\nu \mathbf{k}} \right) \omega^2 \right) \tag{2.7}$$

From the commutation relations between coordinates and momenta, we obtain the following commutation relations

$$\left[\mathrm{H}, \hat{q}^{\pm}_{\nu \mathbf{k}} \right] = -i \hat{p}^{+}_{\nu \mathbf{k}},$$
$$\left[\mathrm{H}, \hat{p}^{\pm}_{\nu \mathbf{k}} \right] = i \omega^2 \hat{q}^{+}_{\nu \mathbf{k}}.$$

Then, we have commutators

$$\left[\mathrm{H}, \hat{\mathbf{E}} \right] = -i \sqrt{\frac{2}{V}} \sum_{\nu \mathbf{k}} \omega \varepsilon_{\nu \mathbf{k}} \left\{ \hat{p}^{+}_{\nu \mathbf{k}} \cos \mathbf{kr} + \hat{p}^{-}_{\nu \mathbf{k}} \sin \mathbf{kr} \right\},$$

$$\left[\mathrm{H}, \hat{\mathbf{H}} \right] = i \sqrt{\frac{2}{V}} \sum_{\nu \mathbf{k}} \omega^2 [\mathbf{n}\varepsilon_{\nu \mathbf{k}}] \left\{ -\hat{q}^{+}_{\nu \mathbf{k}} \cos \mathbf{kr} + \hat{q}^{-}_{\nu \mathbf{k}} \sin \mathbf{kr} \right\}. \tag{2.8}$$

We also have the following expressions

$$\mathrm{rot}\hat{\mathbf{H}} = \sqrt{\frac{2}{V}} \sum_{\nu \mathbf{k}} \omega \varepsilon_{\nu \mathbf{k}} \left\{ \hat{p}^{+}_{\nu \mathbf{k}} \cos \mathbf{kr} + \hat{p}^{-}_{\nu \mathbf{k}} \sin \mathbf{kr} \right\},$$

$$\mathrm{rot}\hat{\mathbf{E}} = \sqrt{\frac{2}{V}} \sum_{\nu \mathbf{k}} \omega^2 [\mathbf{n}\varepsilon_{\nu \mathbf{k}}] \left\{ -\hat{q}^{+}_{\nu \mathbf{k}} \cos \mathbf{kr} + \hat{q}^{-}_{\nu \mathbf{k}} \sin \mathbf{kr} \right\}. \tag{2.9}$$

Note that the last expressions are obtained in the most general form, regardless of the order of the equations used. It can be seen from the expressions (2.8) and (2.9) that

$$\left[\mathrm{H}, \hat{\mathbf{E}} \right] = -i \mathrm{rot}\hat{\mathbf{H}}, \quad \left[\mathrm{H}, \hat{\mathbf{H}} \right] = i \mathrm{rot}\hat{\mathbf{E}}.$$

If the Heisenberg Eq. (2.5) is valid, then we obtain the usual Maxwell equations, i.e.

$$\frac{\partial \hat{\mathbf{E}}}{\partial t} = rot\hat{\mathbf{H}},$$
$$\frac{\partial \hat{\mathbf{H}}}{\partial t} = -rot\hat{\mathbf{E}}.$$

To obtain the fractional Maxwell equations, it is necessary to modify the Heisenberg equation, which in the usual form turns out, as we see, to be not quite appropriate. Thus, in the case of electromagnetic waves in "fractal" media, it is necessary to use the fractional Heisenberg equation for their quantization, i.e.

$$-i\partial^{\alpha}_{0\tau}\hat{F} = [\mathrm{H}, \hat{F}]. \tag{2.10}$$

We note one property of such a Heisenberg equation. It is known that the conserved (time-independent) quantities commute with the Hamiltonian, i.e. The right-hand side of Eq. (2.10) vanishes for such quantities. But on the other hand, the fractional derivative of a constant, i.e. the left-hand side of Eq. (2.10) does not vanish. This means that in this case it is pointless to talk about the conserved quantities. Those we again come to the field dissipation in the medium.

2.3 *Transition to the Schrödinger Picture*

Unlike the Heisenberg representation, in the Schrödinger representation the operators do not depend explicitly on time, and the state vectors in the first case do not depend on time, but in the second they depend. Usually the transition from the Heisenberg representation is carried out by a unitary transformation of the operator and the state vector, i.e. in the Schrödinger representation

$$\hat{F}_s = U\hat{F}U^{+}, \quad \Phi_s = U\Phi.$$

Then

$$-i\frac{\partial \hat{F}_s}{\partial t} = -[\mathrm{H}, \hat{F}_s] + [\mathrm{H}, \hat{F}_s] = 0.$$

In this case the operator is time-independent. For the state vector we have

$$i\frac{\partial \Phi_s}{\partial t} = i\frac{\partial U}{\partial t}\Phi = \mathrm{H}U\Phi = \mathrm{H}\Phi_s,$$

those, we arrive at the Schrodinger equation.

It is obviously impossible to carry out the same operations using such a unitary transformation with fractional derivatives. This is indicated by the fact that the fractional derivative of the product is not calculated for ordinary derivatives, but is an infinite series [36]. It is this fact that is very important, as we see, in the transition from one representation to another. Therefore, to assert that the fractional Schrödinger equation has the form of the ordinary Schrödinger equation with the derivative of the time derivative of the fractional derivative, it is impossible. Usually, use the fractional Schrödinger equation in the form [37]

$$i\partial_{0t}^{\alpha}\Phi_s = \mathrm{H}\Phi_s.$$

But, as we have already said, this equation is generally incorrect in the case of the media considered in this paper; It does not lead to the fractional Heisenberg equations, which we obtained above independently.

2.4 The Schrödinger Equation for Field Operators

All of the above suggests the idea that the starting point in considering such "fractal" quantum systems is the fractional Heisenberg equation in the form in which we recorded it. Because we do not yet know how the fractional Schrödinger equation looks like, then we can not operate it in the form in which it is used in the literature. But we can derive, using the fractional Heisenberg equation, the Schrödinger equation for field operators, which is widely used in quantum statistics when considering many-particle systems [38].

Let us consider a many-particle system with the presence of an interaction between the particles and the external field. The Hamiltonian of such a system in the representation of the second quantization is written as follows [38]

$$\mathrm{H} = \mathrm{H}^{(0)} + V^{(1)} + V^{(2)} + \ldots,$$

where

$$\mathrm{H}^{(0)} = -\frac{1}{2m}\int \hat{\psi}^{+}\Delta\hat{\psi}d^3r - \mu N,$$
$$V^{(1)} = \int \hat{\psi}^{+}U^{(1)}\hat{\psi}d^3r,$$
$$V^{(2)} = \int \hat{\psi}^{+}\hat{\psi}'U^{(2)}\hat{\psi}\hat{\psi}'d^3rd^3r',$$

where $\mathrm{H}^{(0)}$ is the Hamiltonian of free particles, $V^{(1)}$ the operator of their interaction with the external field, $U^{(1)}$, $V^{(2)}$—the operator of interaction of particles with each other, ..., μ is the chemical potential.

The fractional Heisenberg equation for field operators has the form

$$-i\partial_{0\tau}^{\alpha}\hat{\psi} = [\mathrm{H}, \hat{\psi}].$$

Substituting the value instead of the Hamiltonian, we obtain the following, the fractional Schrödinger equation for field operators

$$-i\partial_{0\tau}^{\alpha}\hat{\psi} = \left(-\frac{1}{2m}\Delta - \mu + U^{(1)}\right)\hat{\psi} + \int \hat{\psi}'^{+}U^{(2)}\hat{\psi}'d^3r'\hat{\psi} + \ldots. \tag{2.11}$$

A very important method of quantum field theory, which is used in quantum statistics, is the Green's function method [38, 39]. Therefore, the generalization of the equation for Green's functions to the case of "fractal" systems is of great interest. Using the fractional Heisenberg equation, one can obtain an equation for the Green's function

$$\left(i\partial_{0\tau}^{\alpha} + \frac{\nabla_1^2}{2m} - U(1)\right)G\left(1, 1', U\right) = \delta(1 - 1') \pm i \int dt_2 dr_2 V(1 - 2) G_2(12, 1'2', U),$$

where 1 denotes the aggregate of variables $r_1\tau_1$, 2 denotes $r_2\tau_2$ etc., $G\left(1, 1', U\right)$ is the one-particle Green's function, $G_2(12, 1'2', U)$ is the two-particle Green's function, the sign "±" is chosen depending on the type of particles of the system.

3 The Riesz Fractional Derivative with Respect to the Spatial Coordinate in the Equation for the Green's Function

Fractional derivatives in equations written in space-time form in the transition to Fourier transforms become power laws with a fractional value of degree. In particular, this leads to a fractional-power dispersion law. The idea of the fractional-power dependence of energy on momentum is at the heart of [28, 29], where the question of the possible method of introducing and physical interpretation of fractional integro-differentiation in quantum statistical physics was considered in detail. This idea is demonstrated here. The starting point of the research is the equation for the one-particle Green's function, written in fractional derivatives with respect to the spatial coordinates in the form

$$\left[i\hbar\frac{\partial}{\partial t_1} + E_0\frac{\hbar^{\alpha}}{p_0^{\alpha}}\mathbf{\Delta}_1^{\alpha/2}\right]G(1, 1') = \delta(1 - 1'), \tag{3.1}$$

where 1 means $r_1 t_1$, E_0, and p_0—are the characteristic energy and momentum of the system, $\mathbf{\Delta}_1^{\alpha/2}$—is the fractional Riesz derivative, defined in the one-dimensional case as [40]

$$\Delta_x^{\alpha/2} f(x) = \frac{1}{\Gamma(2-\alpha)\cos\left(\frac{\pi}{2}(2-\alpha)\right)}\frac{\partial^2}{\partial x^2}\int\limits_{-\infty}^{\infty}\frac{f(\xi)d\xi}{|x-\xi|^{\alpha-1}}, \tag{3.2}$$

where $1 < \alpha \leq 2$. The possibility of a nonstrict bound on top of the values taken by the parameter was proved in [41]. The spectrum of the system described by Eq. (3.1), as is easily shown, has the form.

Here we present the results of an investigation of the statistical properties of systems possessing the energy spectrum of a fractional-power type (3.2). The results of this section are interesting for several reasons. First, here we discuss the physical conditions that correspond to the appearance in the equation for the Green's function of the fractional Riesz operator. This will allow us to make certain conclusions to researchers using equations in such operators, to some extent, formally. Secondly, some important properties of systems with a power spectrum of a fractional-power type are considered. The latter gives important information about the connection between the phenomenological parameter-the order of the fractional derivative with the natural parameters of the real statistical system (the interparticle interaction potential, the van der Waals constant, the damping of excited states, etc.).

Let us discuss the possibility of the appearance in the equation for the Green's function of a fractional Riesz operator. Let us try to determine the explicit form of the interparticle interaction potential in the many-particle system, leading to such an operator in the basic equation. The general equation for the Green's function in the presence of an interaction between particles is written as follows [39]

$$\left(i\hbar\frac{\partial}{\partial t_1}+E_0\frac{\hbar^2}{p_0^2}\boldsymbol{\Delta}_1\right)G(1;1')-\int\limits_0^{-i\beta}d\bar{t}_1d\bar{r}_1\sum(r_1t_1;\bar{r}_1\bar{t}_1)G(\bar{r}_1\bar{t}_1;r_{1'}t_{1'})=\delta(1-1'),\tag{3.3}$$

where there $\sum(1;1')$ is a mass operator that contains all information about the interaction of particles with each other. Applying the Fourier transform to both sides of Eq. (3.3), we obtain where there is a mass operator that contains all information about the interaction of particles with each other. Applying the Fourier transformation to both sides of the Eq. (3.3), we obtain where is the mass operator that contains all the information about the interaction of particles with each other. Applying the Fourier transform to both sides of Eq. (3.3), we obtain

$$\left[\omega-\frac{E_0}{p_0^2}p^2-\sum(p,\omega)\right]G(p,\omega)=1.\tag{3.4}$$

The spectrum of the system, according to (3.4), is defined as $\omega(p)=\left(E_0/p_0^2\right)p^2+\mathrm{Re}\sum(p,\omega)$. The spectrum (2.2), as is easily seen, corresponds to the mass operator

$$\sum(p)=E_0\left(p_0^{-\alpha}|p|^\alpha-p_0^{-2}p^2\right).$$

On the other hand, the mass operator, in general, is defined by the expression [39]

$$\sum(p,\omega)G(p,\omega)=i\int drdt e^{-i\frac{\omega t}{\hbar}+\frac{ipr}{\hbar}}\int dr_2V(r_1-r_2)G_2(12,1'2')_{t_1=t_2}.\tag{3.5}$$

or

$$\frac{E_0\left(p_0^{-\alpha}|p|^\alpha - p_0^{-2}p^2\right)}{\omega - E_0 p_0^{-\alpha}|p|^\alpha} = i\int drdte^{-i\frac{\omega t}{\hbar}+\frac{ipr}{\hbar}}\int dr_2 V(r_1 - r_2)G_2(12, 1'2')_{t_1=t_2}, \tag{3.6}$$

there $G_2(12, 1'2')$ is a two-particle Green's function, $V(r)$ is the interparticle interaction potential. Expression (3.6) determines those conditions, i.e. $V(r)$ the potential $V(r)$ that leads to the spectrum (2.2). However, it is impossible to solve this exact equation in the general case. We use the approximation for the mass operator. It can be shown that the mass operator given above for the system under consideration does not contain collisions of particles. Therefore, we confine ourselves to approximations of lower order. In the Hartree-Fock approximation for a fermion gas we have

$$\sum_{HF}(p) = n\int V(r)dr - \int \frac{dp'}{(2\pi)^3}\frac{V(p-p')}{\exp\left(\beta\left[E_0 p_0^{-\alpha}|p'|^\alpha - \mu\right]\right)+1},$$

where $\beta = 1/kT$, μ is the chemical potential, is the particle concentration in the translationally invariant system. It is obvious that the dependence of the energy on the pulse can be modified only at the expense of the Fock part of the mass operator. With this in mind, we write

$$E_0\left(p_0^{-\alpha}|p|^\alpha - p_0^{-2}p^2\right) = -\int \frac{d^3p'}{(2\pi\hbar)^3}\frac{V(p-p')}{e^{\beta\left[E_0 p_0^{-\alpha}|p'|^\alpha - \mu\right]}+1}, \tag{3.7}$$

The spectrum of such a system has the form $\omega(p) = E_0 p_0^{-\alpha}|p|^\alpha + n\int V(r)dr$. Such a spectrum corresponds to an equation for the Green's function

$$\left[\omega - E_0 p_0^{-\alpha}|p|^\alpha - n\int V(r)dr\right]G(1, 1') = \delta(1-1'),$$

which after the Fourier transformation gives the following equation

$$\left[i\hbar\frac{\partial}{\partial t_1} + E_0\frac{\hbar^\alpha}{p_0^\alpha}\Delta_1^{\alpha/2} + \int V(r_1 - r_2)\langle n(r_2)\rangle dr_2\right]G(1, 1') = \delta(1-1'),$$

We note that this equation differs from (3.1) by the presence of a Hartree term. In [28], the Hartree term in the equation for the Green's function was neglected. In this paper, we will also follow this approximation, assuming, as usual, the presence of a static background creating an equal field of the opposite sign.

We now turn to the clarification of the conditions that lead to the spectrum (2.2). For this it is necessary to solve the integral Eq. (3.7) with respect to the potential V. In other words, we solve the inverse spectral problem. Applying the transformation to both sides of Eq. (3.7) of Fourier, we obtain

$$V(r) = \frac{E_0 \int \frac{d^D k}{(2\pi)^D} \left(k_0^{-2} k^2 - k_0^{-\alpha} |k|^\alpha\right) e^{-ikr}}{\int \frac{d^D k}{(2\pi)^D} \frac{e^{-ikr}}{e^{\beta\left[E_0 k_0^{-\alpha} |k|^\alpha - \mu\right]} + 1}}, \tag{3.8}$$

where we went to the variables $k = p/\hbar$, $V(r) = \int d^D k/(2\pi)^D V(k) e^{-ikr}$, D is the dimension of the system $D = 1, 2, 3$ under consideration. To determine the integral $\int |p|^\alpha e^{-ip\xi} p^{D-1} dp$, we use the usual method.

$$\int |p|^\alpha e^{-ip\xi} p^{D-1} dp = \int\limits_0^\infty p^{\alpha+D-1} \left(e^{ip\xi} + e^{-ip\xi}\right) dp$$

$$= \lim_{\delta \to 0} \int\limits_0^\infty p^{\alpha+D-1} \left(e^{ip(\xi + i\delta)} + e^{-ip(\xi - i\delta)}\right) dp.$$

Then for $D = 3$

$$V(r) = \frac{E_0 \left(k_0^{-2} \frac{\Gamma(5)}{r^5} - k_0^{-\alpha} \frac{\Gamma(\alpha+D)}{r^{\alpha+3}}\right)}{\int \frac{d^3 k}{(2\pi)^3} \frac{e^{-ikr}}{e^{\beta\left[E_0 k_0^{-\alpha} |k|^\alpha - \mu\right]} + 1}}$$

The integral in the denominator, which we denote as $K(r)$ not being computed in general form. Therefore, consider the case of low density, when $e^{-\beta\mu} \gg 1$. Then

$$K(r) = e^{\beta\mu} \int \frac{d^3 k}{(2\pi)^3} e^{-\beta E_0 k_0^{-\alpha} |k|^\alpha} e^{-ikr}.$$

In the very rough approximation (in general, this is true only for $\alpha \to 2$), we put this integral equal $K(r) = A e^{\beta\mu} e^{-\frac{1}{\beta E_0 k_0^{-\alpha}} |r|^\alpha}$, where A is the constant appearing in the integration. Thus, for the interaction potential (3.8) we obtain

$$V(r) = \frac{E_0}{A} e^{-\beta\mu} e^{\frac{1}{\beta E_0 k_0^{-\alpha}} |r|^\alpha} \left(k_0^{-2} \frac{\Gamma(5)}{r^5} - k_0^{-\alpha} \frac{\Gamma(\alpha+3)}{r^{\alpha+3}}\right) \tag{3.9}$$

We have obtained the potential that, in this approach, leads to a fractional derivative of Riesz in the coordinate. At $\alpha = 2$ this potential is zero. Let the system under consideration be at a temperature $T \sim 10$ °K. Let the mass of particles be on the order of the proton mass $M \approx 10^{-28}$ Kg. We take the characteristic energy equal to $E_0 = 0.1$ eV, and for the characteristic wave number, according to the formula $E_0 = k_0^2 \hbar^2 / 2M$, we obtain $k_0 \approx 10^{10} \mathrm{M}^{-1}$. In addition, suppose that $e^{-\beta\mu} \approx 20$. Then, the potential obtained in the case of a three-dimensional system behaves as shown in the figure.

In terms of its physical meaning, the derivative of a fractional order in coordinate means the inclusion of spatial correlations. As can be seen in Fig. 1, Indeed, the transition to fractional derivatives is equivalent to introducing a pairwise interpar-

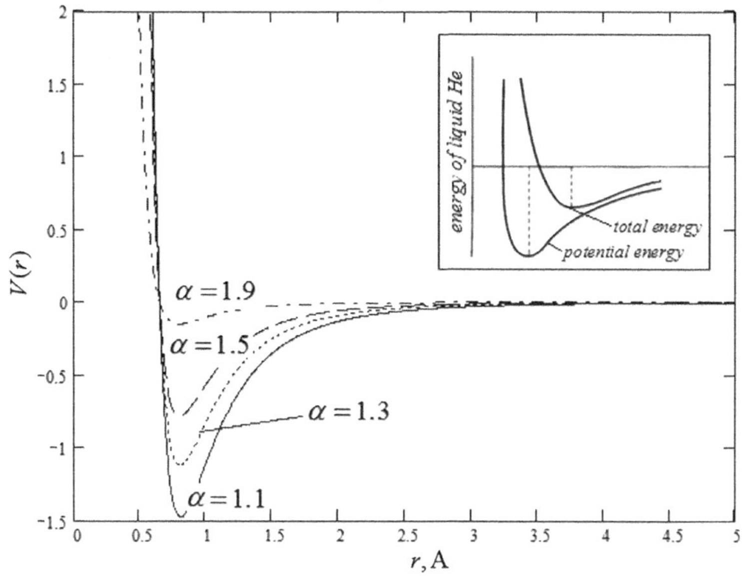

Fig. 1 The interaction potential (3.9) for different values. For comparison, the small figure shows the potential for the interatomic interaction of liquid helium (see, for example, [42])

ticle potential, leading to a nonideality of the gas. However, unlike the traditional interparticle potential, which is purely potential in nature and does not depend on the momentum, in this case we have a fundamentally different case. The point is that the transition to a fractional derivative, leading to a fractional spectrum, means taking into account the effective interaction between the particles, and, what is extremely important, this interaction is actually implicitly dependent on the momentum. In this connection we note that the consideration of the case in which the energy of interaction between particles depends on the momentum in traditional statistical physics is impossible in principle. Thus, the transition to fractional derivatives with respect to the coordinate, in fact, means taking into account the dependence of the interaction energy on the pulse, and thereby broadens the limits of applicability of statistical physics, allowing also to take into account the dependence of the interaction energy on the momenta.

Let us analyze this potential using the example of a system of charged particles. In such a system, the interaction potential between particles is, in general, determined according to the expression [43]

$$V(r) = \int \frac{e^2 e^{-ikr} dk}{k^2 \varepsilon(k)}.$$

The quantity e^2/k^2 is the Fourier transform of the Coulomb interaction in a three-dimensional system and $\varepsilon(k)$ is the permittivity in a medium with spatial dispersion.

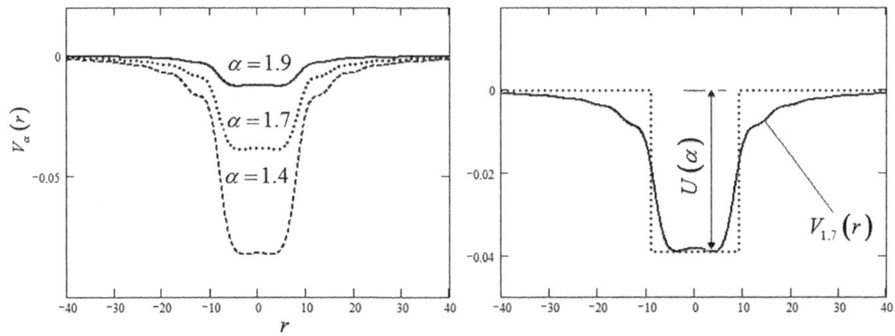

Fig. 2 Interparticle interaction potential in a degenerate electron gas, which leads to the energy spectrum of a fractional-power type formula after Eq. (3.7)

In a medium without dispersion, we have $\varepsilon = const$, and therefore $V(r) = q^2/\varepsilon r$, we arrive at the Coulomb law. If the permittivity is of the form $\varepsilon(k) = 1 + k_0^2/k^2$, then we arrive at the Debye screening $V(r) = q^2 e^{-k_0 r}/r$, where k_0^{-1} is the Debye radius. It is of interest to solve equation

$$V_\alpha(r) = \int \frac{q^2 \mathrm{e}^{-ikr} dk}{k^2 \varepsilon_\alpha(k)},$$

where $V_\alpha(r)$ is the expression (3.9), relative $\varepsilon_\alpha(k)$. In this case it can be asserted that in the system of charged particles with a dielectric permittivity $\varepsilon_\alpha(k)$ the interparticle potential is (3.9), which leads to a fractional-power spectrum, and, consequently, to the fractional Riesz derivative in the Green's function equation.

Let us consider the case of a degenerate electron gas. In this case it is necessary to put

$$\left[\exp\left(\beta\left[E_0 p_0^{-\alpha}|p'|^\alpha - \mu\right]\right) + 1\right]^{-1} = \Theta\left(E_F - E_0 p_0^{-\alpha}|p|^\alpha\right).$$

In addition, we set $E_0 = E_F$ and $p_0 = p_F$. Taking this into account, we obtain from (3.8)

$$V_\alpha(r) = E_F \mathrm{Re}\left\{ \frac{\int_0^{k_F} k^2 dk \left(k_F^{-2} k^2 - k_F^{-\alpha} k^\alpha\right) \mathrm{e}^{-ikr}}{\int_0^{k_F} k^2 dk \mathrm{e}^{-ikr}} \right\}, \tag{3.10}$$

In Fig. 2, the dependence of the interparticle potential (3.10) on the distance between the particles is shown. Energy is given in units of Fermi energy, and distance in angstroms.

The interaction potential, as is easily seen from the figure, is a potential well whose depth depends on the parameter α.

The interaction potential, as is easily seen from the figure, is a potential well whose depth depends on the parameter.

Another important conclusion is that the interaction in this case is of the nature of attraction, and therefore it can be said that we are dealing with a phenomenon analogous to the phenomenon of superconductivity, in which electrons, due to the exchange of phonons, begin to attract and form Cooper pairs. In addition, in this case, the intensity of the electron-electron attraction is a function of the parameter α. Apparently, the following effect should be realized in such a system. At a certain temperature in the system under consideration, the electrons are localized in the potential well shown in the figure. The transition to such a state is called an f-transition. The temperature of the f-transition is obviously determined by the depth of the well, which in turn depends on the parameter α. Thus, for the transition temperature we have $U(\alpha) = k_B T_f$. Let us make some estimates. For the value $\alpha = 1.9$, the depth of the well is of the order $U(\alpha = 1.9) = 0.01$ эВ Hence we find that for this case $T_f \approx 100K$. Similarly, for a value $\alpha = 1.7$, for example, we get $T_f \approx 400K$.

We consider the question of the equation of the state of a gas (for definiteness we take a Fermi gas) having the spectrum (2.2), and, respectively, described by Eq. (3.1). The Fermi distribution for such a gas has the form $n_\alpha(\beta, p) = (\exp[\beta(|p|^\alpha - \mu)] + 1)^{-1}$, where a system of units is used, in which the constants that are unessential at this stage are unity, i.e. $E_0 = 1$, $p_0 = 1$. The equation of state for a system with such a distribution was obtained in [28]. In this work, to derive the equation of state, the relation was used $dP = nd\mu$, where P is the pressure. Then it was differentiated $n_\alpha(\beta, p)$ to find the differential of the chemical potential. The main consideration used in [28] when performing such actions on a function $n_\alpha(\beta, p)$ is as follows. The parameter α implicitly contains an effective interaction between the particles. This interaction, naturally, is determined by the concentration of the particles. Therefore, when differentiating, one should keep in mind that the parameter α is not a constant value. This parameter is a density functional. This means that this parameter also requires differentiation. In a sense, these considerations are very similar to similar statements of the Fermi-liquid theory. Indeed, in the latter case the distribution is written as $n(\beta, p) = (\exp[\beta(\varepsilon(p) - \mu)] + 1)^{-1}$ where $\varepsilon(p)$ is the quasiparticle energy, which in turn is a functional of the density, i.e. $\varepsilon = \varepsilon(n, p)$. With the help of such considerations, a state equation for a Fermi system possessing an energy spectrum of a fractional-power type

$$P = \frac{1}{\beta} n + 2 \frac{1}{\beta} e^{\beta\mu} \int \frac{d^3 p}{(2\pi)^3} \left(1 - e^{\beta(p^2 - |p|^\alpha)}\right) e^{-\beta|p|^\alpha}, \tag{3.11}$$

where n is the concentration of the ideal gas (at $\alpha = 2$). It is not difficult, with the help of the same arguments, to obtain the equation of state for a Bose gas.

Equation (3.11) shows that the power law of dispersion leads, in the final analysis, to the appearance of some additional pressure in the equation of state, that is, in fact, takes into account the interaction between the particles. The distribution function that determines this additional pressure is essentially the Weibull distribution (the stretched exponent), which is closely related to the fractal properties of the system

[44]. This additional pressure is zero when u and we arrive at the equation of state of an ideal gas.

Let us investigate Eq. (3.11). As it was said, this equation describes the gas of interacting fermions. Let us try to make some general remarks on the nature of this interaction through more familiar approaches. Equation (3.11), written in the form

$$P - I_\alpha = \frac{N}{V}kT, \tag{3.12}$$

where $I_\alpha = 2\beta^{-1}\mathrm{e}^{\beta\mu}\int d^3p/(2\pi)^3\left(1-\mathrm{e}^{-\beta(p^2-|p|^\alpha)}\right)\mathrm{e}^{-\beta|p|^\alpha}$ it is very similar to the Van der Waals equation, which, however, does not contain information on the finiteness of the particle sizes. This is due to the fact that the approximation considered does not take into account the collisions and the fact that the particles can not penetrate into each other. If we neglect the finite particle size effects, then the van der Waals equation can be written as

$$P - n^2a = \frac{N}{V}kT,$$

Comparing this equation with Eq. (3.12), we obtain

$$n^2a = 2\frac{1}{\beta}\mathrm{e}^{\beta\mu}\int\frac{d^3p}{(2\pi)^3}\left(1-\mathrm{e}^{-\beta(p^2-|p|^\alpha)}\right)\mathrm{e}^{-\beta|p|^\alpha}.$$

In order to obtain an expression for the chemical potential, we resort to an approximation, namely, we use the expression for the chemical potential of a Boltzmann ideal gas. This approximation is justified in many cases [45]. The chemical potential of a Boltzmann gas is, to within insignificant constants, a value equal to $\mu = \beta^{-1}\ln\left[n\beta^{3/2}\right]$ [45]. Then we have

$$na = 2\beta^{1/2}\int\frac{d^3p}{(2\pi)^3}\left(1-\mathrm{e}^{-\beta(p^2-|p|^\alpha)}\right)\mathrm{e}^{-\beta|p|^\alpha}.$$

For specific calculations, it is necessary to go to dimensional quantities. In dimension variables, the integral is written as

$$I_\alpha = 2\frac{1}{\beta}\mathrm{e}^{\beta\mu}\int\frac{d^3k}{(2\pi)^3}\left(1-\mathrm{e}^{-\beta E_0\left(k_0^{-2}k^2-k_0^{-\alpha}|k|^\alpha\right)}\right)\mathrm{e}^{-\beta E_0k_0^{-\alpha}|k|^\alpha},$$

where $k = p/\hbar$. The chemical potential of a Boltzmann gas in dimensional variables has the form $\mu = \beta^{-1}\ln\left[n/2\left(2\pi\hbar^2m^{-1}\beta\right)^{3/2}\right]$ [45]. For the Vander Waals constant, therefore, we obtain the following expression relating this constant to the parameter α

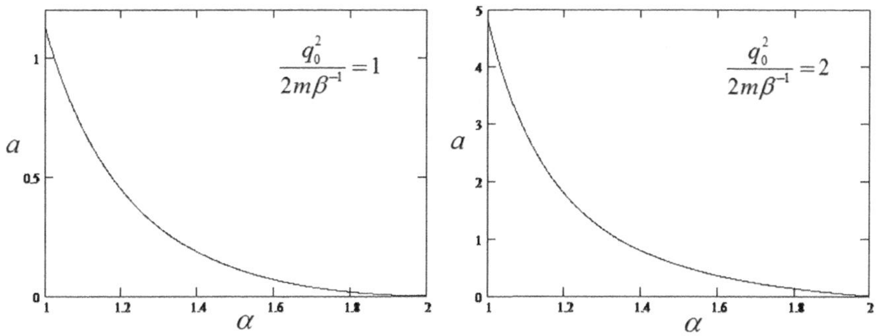

Fig. 3 Dependence of the van der Waals constant a on the parameter α. When constructing curves, it is conventionally assumed that $\beta^{-1}/n(\pi)^{3/2} = 1$

$$a = \frac{\beta^{1/2}}{n}\left(\frac{2\pi\hbar^2}{m}\right)^{3/2} \times \int \frac{d^3k}{(2\pi)^3}\left(1 - e^{-\beta E_0\left(k_0^{-2}k^2 - k_0^{-\alpha}|k|^{\alpha}\right)}\right)e^{-\beta E_0 k_0^{-\alpha}|k|^{\alpha}} \tag{3.13}$$

This ratio is valuable from the point of view of the possibility of describing real systems by the method of fractional integro-differentiation, which is carried out with the help of Eq. (3.1). In Fig. 3, this dependence is shown.

Each value of the van der Waals constant is associated with a single value of the parameter α. Those systems with different ones can be described by the equation of state (3.12) with a parameter α corresponding to this a.

Let us consider with the same approach the problem of electron-phonon interaction. The model that takes into account the electron-phonon interaction is constructed on the basis of the quadratic dependence of the energy on the momentum and gives the interaction amplitude for the interaction amplitude [46]

$$\Phi(q) \sim q^{1/2}.$$

However, this model does not take into account all the details and various mechanisms of electron-phonon interaction [47–49] and, in order to improve the coincidences with the experimental data, modify this function, namely, change the degree of dependence on the pulse already at a purely phenomenological level [46]. It is of interest to consider the fractional-differential model of the electron-phonon interaction. In doing so, we use the elementary approach to this problem, carried out in [46].

Consider a system of interacting ions and electrons. The ions are located at the nodes of the crystal lattice, and the electrons are considered to be free to a certain degree of accuracy. The starting point of the discussion is the fractional equation for the Green's function (1). As was said above, the systems described by such equations possess the spectrum formula after Eq. (3.7). Taking this into account, the total energy for such systems can be written in the form

$$E = |p|^{\alpha} + U(r) \tag{3.14}$$

$U(r)$—the potential energy of ions, or in dimensional variables

$$E = \frac{q_0^{2-\alpha}}{2M}|q|^{\alpha} + U(r) \tag{3.15}$$

Usually, ions are represented as harmonic oscillators. In this case, when the dependence on the pulse becomes fractional power, it is more likely to expect a deviation from the harmonic law. In connection with this, we consider this problem in a general form, following [21], where a class of Hamiltonians of the type

$$H = |p|^{\alpha}|r|^{\alpha'} + |r|^{\beta}|p|^{\beta'}$$

Consider the case when $\alpha' = \beta' = 0$. Then we have the following Hamiltonian

$$H = |p|^{\alpha} + |r|^{\beta}$$

Taking this into account, we rewrite the expression for the total energy in the form

$$E = \frac{q_0^{2-\alpha}}{2M}|q|^{\alpha} + \frac{k}{2}|r|^{\beta} \tag{3.16}$$

where k is the elastic constant. With allowance for (3.16), the total energy of the wave of deviations of the lattice sites from the equilibrium position is represented in the form

$$E = \int \frac{q_0^{2-\alpha}}{2M} n \left| M \frac{\partial \xi_q}{\partial t} \right|^{\alpha} dr + \int \frac{k}{2} |\Delta \xi_q|^{\beta} dr$$

The wave of deviations of the nodes $\boldsymbol{\Delta}\xi_q$ from the equilibrium position is related to the gradient by the relation

$$\Delta \xi_q = \xi_q(r+a) - \xi_q(r) = a \frac{\partial \xi_q}{\partial r}$$

With this in mind, we finally obtain

$$E = \frac{q_0^{2-\alpha}}{2} n M^{\alpha-1} \int \left| \frac{\partial \xi_q}{\partial t} \right|^{\alpha} dr + \int \frac{\chi}{2} \left| \frac{\partial \xi_q}{\partial r} \right|^{\beta} dr \tag{3.17}$$

If we represent a wave of deviations in the form of a standard exponent

$$\xi_q = A b_q^{+} \mathrm{e}^{-i\omega t + \frac{i}{\hbar} q r}$$

where A is the amplitude of the deviations, b_q^+ is the phonon production operator, then

$$E = \frac{q_0^{2-\alpha}}{2} n M^{\alpha-1} \left(A b_q^+\right)^{\alpha} \omega^{\alpha} V + \frac{\chi}{2}\left(\frac{q}{\hbar}\right)^{\beta} \left(A b_q^+\right)^{\beta} V$$

Because the phonon energy $\hbar\omega$ is, and also $\hbar\omega = cq$ and $c^2 = \chi/nM$, then we obtain an expression that determines the dependence of the amplitude on the momentum

$$\begin{aligned} cq = {} & \frac{q_0^{2-\alpha}}{2} n M^{\alpha-1} \left(A b_q^+\right)^{\alpha} (cq)^{\alpha} V \\ & + \frac{c^2 n M}{2}\left(\frac{q}{\hbar}\right)^{\beta} \left(A b_q^+\right)^{\beta} V \end{aligned} \qquad (3.18)$$

In the case of a quadratic Hamiltonian, i.e. $\alpha = \beta = 2$ when we obtain a known result [46]

$$A(q) = \gamma \frac{1}{\sqrt{nMcqV}},$$

where $\gamma = \hbar/b_q^+$.

Let us consider the case $\alpha = \beta$. In this case, from (3.18) we obtain a more general dependence of the interaction amplitude on the momentum

$$A(q) = \gamma' \left(M n V c^{\alpha-1}\right)^{-1/\alpha} q^{\frac{1-\alpha}{\alpha}},$$

where $\gamma' = \left[\frac{1}{2}\left(q_0^{2-\alpha} M^{\alpha-2} + c^{2-\alpha}\right)\right]^{-1/\alpha} \frac{\hbar}{b_q^+}$.

The amplitude of the electron-phonon interaction is determined by the amplitude of the wave of deviation ξ_q of the perturbed density of the lattice sites $n_q(r,t)$ from the equilibrium value n, $n_q(r,t) = n(r,t) - n \ll n$. The latter is proportional to the gradient $\partial\xi_q/\partial r$. Then we get

$$n_q(r,t) = const\left(M n V c^{\alpha-1}\right)^{-1/\alpha} q^{\frac{1}{\alpha}} \left(b_q^+ \exp\left(-i\omega t + \frac{i}{\hbar} q r\right) + k.c.\right)$$

Taking this expression into account for the Hamiltonian of the interaction of the electron gas with the resulting lattice vibration,

$$H_{ef} = \sum_{pq} const \cdot q^{\frac{1}{\alpha}} \left(a_{p-q}^+ a_p b_q^+ + a_{p+q}^+ a_p b_q\right)$$

which, supplemented by the Hamiltonian of free particles, is a well-known expression of the Frohlich Hamiltonian, but with the amplitude of the electron-phonon interaction

$$\Phi(q) \sim q^{1/\alpha}. \tag{3.19}$$

Expression (3.18) is a wide class of dependencies $A(q)$, which allows one to interpret the results of real experiments by varying only with a parameter α.

Different models of electron-phonon interaction are particularly interesting in the theory of superconductivity. Indeed, the amplitude of the electron-phonon interaction is an important parameter that determines the critical temperature of the transition to the superconducting state. It is known that the classical Bardeen-Cooper-Schrieffer (BCS) superconductivity theory, being an idealized model, can not adequately describe the properties of all conventional superconductors, allowing serious errors, for example, in calculating magnesium diboride parameters-an unusual compound having a very high transition temperature. The problem is that the BCS does not extend to the case of a strong electron-phonon interaction. The "improved" version of the BCS, proposed by Gerasim Eliashberg, allows obtaining more accurate formulas for calculating the critical temperature. For example, physicists solved the Eliashberg equations immediately in two exotic cases: for pure and for doped graphene. Since in pure graphene the Fermi level lies at the Dirac point [50], here one must speak of a "multi-zone" pairing, in which particles from both bands participate [51]. Doping graphene implies a much more effective pairing, in which particles belonging to only one zone participate. In connection with the foregoing, we can conclude that in the case of graphene and other objects where the conditions for high-temperature superconductivity are realized, it is of interest to apply the fractional differential model of superconductivity, the parameter of which will be the more general expression (2.19), from which it follows that the amplitude of the electron-phonon interaction at more (and in the case, the maximum) than in the case. Some attempts to describe high-temperature superconductivity using the spectrum formula after Eq. (3.7) were undertaken in Refs [12, 52].

Thus, taking this and the previous section into account, we can write the following general fractional equation for the Green's function

$$\begin{aligned}&\left(\frac{i}{t_0}\partial^{\alpha}_{0\tau} + E_0\frac{\hbar^{\beta}}{p_0^{\beta}}\Delta_1^{\beta/2} - U(1)\right)G\big(1, 1', U\big)\\&= \delta(1-1') \pm i\int dt_2 dr_2 V(1-2) G_2(12, 1'2', U)\end{aligned} \tag{3.20}$$

4 Fractional Analysis of Instability in a Gas Discharge

Progress in the study of the gas discharge physics under high pressure is largely determined by knowledge of the physical properties of the discharge. Especially, this corresponds the initial stage, which in gases and liquids is accompanied by the generation and propagation of specific ionization waves. The wide practical application of gas discharge various forms stimulates studies of their spatial structure.

In [53], the initial stage of the development of the ionization wave front instability due to the multiplication of electrons of low background density is considered. An expression for the growth rate of small perturbations is found. It is shown that the propagation front is unstable with respect to small perturbations forming protrusions or dips. The growth rate of the instability can be defined as a function of the reduced field strength that is universal for a given gas.

In [54], the microstructure of the current channel was experimentally detected in the breakdown of homogeneous air gaps by voltage pulses of the nanosecond range in electric fields insufficient to form of a streamer. As a mechanism for the formation of a microstructure, the development of the instability of the ionization process in the avalanche stage is proposed, that leads to the formation of a self-similar spatial structure. It is shown that the microstructure of streamer discharges in homogeneous gaps can also be explained within the framework of the proposed model.

There is a large number of both experimental and theoretical studies on the stability of ionization fronts. Nevertheless, there is no unified theory of the development of instability. The latter is due to the complexity of accounting for all important factors affecting processes. Thus, the development of new effective approaches in this field of research is extremely important.

Finally, we note following. As is well known, the turbulent state is natural for a plasma. In this state, the plasma obeys the laws of anomalous diffusion [55–59]. In such a state, the mean free paths of particles are power functions [59]. This can lead to a fractional equation for the distribution function (details see in [6]).

It should be noted that fractional derivatives cannot be used on fractal since it cannot be considered a linear (vector) space. A mathematically correct approach is given in the books [60, 61].

Here, using an approach based on the kinetic equation of fractional order on the time variable, two types of instability in a gas discharge are investigated: the instability of the electron avalanche and the sticking instability in a nonself-maintained discharge.

4.1 Instability of an Electron Avalanche in a Gas Discharge

A standard approach to investigating the stability of an avalanche is based on the use of the kinetic equation for the electron concentration in the avalanche (an adiabatic approximation is used when ion motion is neglected)

$$\frac{\partial n}{\partial t} = \alpha v_{dr} n - v_{dr} \frac{\partial n}{\partial z} + D_e \Delta n. \tag{4.1}$$

where α is the Townsend ionization coefficient, v_{dr} is the electron drift velocity, D_e is the electron diffusion coefficient, the axis Oz is directed along the field. In Eq. (3.1) we pass to the Riemann-Liouville fractional derivatives [3]

$$\partial_{-\infty x}^{\beta} f(x) = \frac{1}{\Gamma(1-\{\beta\})} \frac{\partial^{[\beta]+1}}{\partial x^{[\beta]+1}} \int\limits_{-\infty}^{x} \frac{f(\xi)d\xi}{(x-\xi)^{\{\beta\}}}. \tag{4.2}$$

where $[\beta]$ is the integer part of β and $0 \leq \{\beta\} < 1$ is the fractional part of β. (Note that the fractional derivative (4.2) is also called the Liouville derivative [19]). Then

$$\frac{1}{t_0}\partial_{-\infty\tau}^{\beta} n = \alpha \upsilon_{dr} n - \upsilon_{dr}\frac{\partial n}{\partial z} + D_e \Delta n, \tag{4.3}$$

where $\tau = t/t_0$ is a dimensionless time, t_0 is the some characteristic time. The problem that solving in the present paper is a rare case when the parameter t_0 is not included in the final result. In general, as t_0 one can use, for example, the characteristic ionization time. We will be sought the solution in a standard way in a following form: $n \propto \exp(-i\omega\tau + ikr)$. Then[1]

$$\frac{1}{t_0}(-i\omega)^{\beta} = \alpha \upsilon_{dr} - i\upsilon_{dr}k_z - D_e k^2, \tag{4.4}$$

where $k = 2\pi/\lambda$, λ is a characteristic size of the disturbance inhomogeneity. Next, we note that

$$(-i\omega)^{\beta} = |\omega|^{\beta} \exp\left(-i\beta\frac{\pi}{2} + i\beta \arctan\frac{\operatorname{Im}\omega}{\operatorname{Re}\omega}\right) \tag{4.5}$$

Then

$$|\omega|^{\beta} \cos\beta\left(\arctan\frac{\operatorname{Im}\omega}{\operatorname{Re}\omega} - \frac{\pi}{2}\right) = t_0\left(\alpha\upsilon_{dr} - D_e k^2\right),$$

$$|\omega|^{\beta} \sin\beta\left(\arctan\frac{\operatorname{Im}\omega}{\operatorname{Re}\omega} - \frac{\pi}{2}\right) = -t_0 \upsilon_{dr} k_z.$$

The last equations give

$$\begin{aligned}\operatorname{Im}\omega = t_0^{\frac{1}{\beta}}\left(\left(\alpha\upsilon_{dr} - D_e k^2\right)^2 + \upsilon_{dr}^2 k_z^2\right)^{\frac{1}{2\beta}} \\ \times \cos\left(\frac{1}{\beta}\arctan\frac{\upsilon_{dr}k_z}{\alpha\upsilon_{dr} - D_e k^2}\right)\end{aligned} \tag{4.6}$$

The condition for the onset of instability is the following inequality

$$\operatorname{Im}\omega \geq 0 \tag{4.7}$$

Then condition (4.7) can be rewritten in the form

[1] We used the property of the fractional Riemann-Liouville derivative: $\partial_{-\infty x}^{\beta} e^{ax} = a^{\beta} e^{ax}$.

$$\left(\left(\alpha\upsilon_{dr} - D_e k^2\right)^2 + \upsilon_{dr}^2 k_z^2\right)^{\frac{1}{2\beta}} \cos\left(\frac{1}{\beta} \arctan \frac{\upsilon_{dr} k_z}{\alpha\upsilon_{dr} - D_e k^2}\right) \geq 0, \tag{4.8}$$

or

$$\cos\left(\frac{1}{\beta} \arctan \frac{\upsilon_{dr} k_z}{\alpha\upsilon_{dr} - D_e k^2}\right) \geq 0 \tag{4.9}$$

or

$$\alpha\upsilon_{dr} - D_e k^2 \geq \frac{\upsilon_{dr} k_z}{\tan \frac{\beta\pi}{2}} \tag{4.10}$$

When $\beta = 1$ we obtain the usual condition

$$\alpha\upsilon_{dr} - D_e k^2 \geq 0. \tag{4.11}$$

Let us find the values of k, under which condition (4.10) is met. Firstly, we note that $k_z = k \cos\theta$. We consider the directed discharge and we neglect reflections back. This means that $0 < \theta < \pi/2$. Then Eq. (4.10) can be rewritten as follow

$$\alpha\upsilon_{dr} - D_e k^2 - \frac{\upsilon_{dr} k \cos\theta}{\tan \frac{\beta\pi}{2}} = 0. \tag{4.12}$$

The solutions of Eq. (4.12) have the form

$$k_{1,2} = \frac{-\upsilon_{dr} \cos\theta \pm \sqrt{D}}{2D_e \tan \frac{\beta\pi}{2}}, \tag{4.13}$$

where

$$D = (\upsilon_{dr} \cos\theta)^2 + 4 \tan^2 \frac{\beta\pi}{2} \alpha\upsilon_{dr} D_e. \tag{4.14}$$

It is clear from (4.13) that the solution when $k_2 < 0$ has no physical meaning. Therefore, this solution should be discarded. Then

$$k \leq \frac{-\upsilon_{dr} \cos\theta + \sqrt{D}}{2D_e \tan \frac{\beta\pi}{2}}. \tag{4.15}$$

For the length λ, we obtain the condition

$$\lambda \geq \frac{4\pi D_e \tan \frac{\beta\pi}{2}}{-\upsilon_{dr} \cos\theta + \sqrt{D}}. \tag{4.16}$$

For $\beta \to 1$ (or any odd number) we have the condition

$$\lambda \geq \frac{2\pi\sqrt{D_e}}{\sqrt{\alpha v_{dr}}}, \tag{4.17}$$

completely coinciding with the well-known [1].

For $\beta \to 2$ (or any even number) we have the condition $\lambda \geq \infty$, i.e. instability vanish completely.

Let us analyze the obtained result. Firstly, we note that condition (4.16), unlike (4.17), is anisotropic with respect to the field direction. In other words, the instability condition that is satisfied for a given set of values of the angle θ will not be satisfied for some other values of this angle. Obviously, such anisotropy occurs when the ionization wave front is not flat. Indeed, in the case of a plane front $\theta = 0$. It is interesting that the anisotropy effect occurs when the equation is used in fractional derivatives and vanishes when $\beta \to 1$. Further, the right-hand side of (4.16) in the angle interval (0, π/2) is a decreasing function of θ. I.e. the critical value of λ, at which an instability arises, is greater along the electric field than across. Finally, the instability condition includes an additional parameter β, but this condition does not contain a constant t_0. The latter is extremely important, because in most cases of using fractional analysis the real meaning of such a parameter is one of the main problems (Fig. 4).

4.2 Adherent Instability in a Non-self-sustaining Discharge

Let us consider the equation for the density of secondary electrons $n_e(t)$ in a discharge controlled by an electron beam.

$$\frac{dn_e}{dt} = S - \beta_r n_e - \alpha n_e^2. \tag{4.18}$$

Here S is the formation rate of secondary electrons during the passage of the primary electron beam, which is assumed to be constant, β_r is the adhesion coefficient (adhesion of electrons to neutral particles), and α is the electron-ion recombination coefficient.

In Eq. (4.18) we pass to the fractional derivatives

$$\frac{1}{t_0}\partial^{\beta}_{-\infty\tau} n_e = S - \beta_r n_e - \alpha n_e^2. \tag{4.19}$$

Let us suppose that in the system there are local fluctuations leading to perturbation δn_e and δE (E—external dielectric field intensity) in some area

$$\begin{cases} n_e \to n_{e0} + \delta n_e \\ E \to E + \delta E. \\ \beta_r(E) \to \beta_r(E + \delta E) \approx \beta_r(E) + (\partial\beta_r/\partial E)\delta E \end{cases} \tag{4.20}$$

Fig. 4 Dependence of the instability parameter: **a** on the order of the fractional derivative for different values of the angle θ, **b** on the angle θ for different values of the order of the fractional derivative

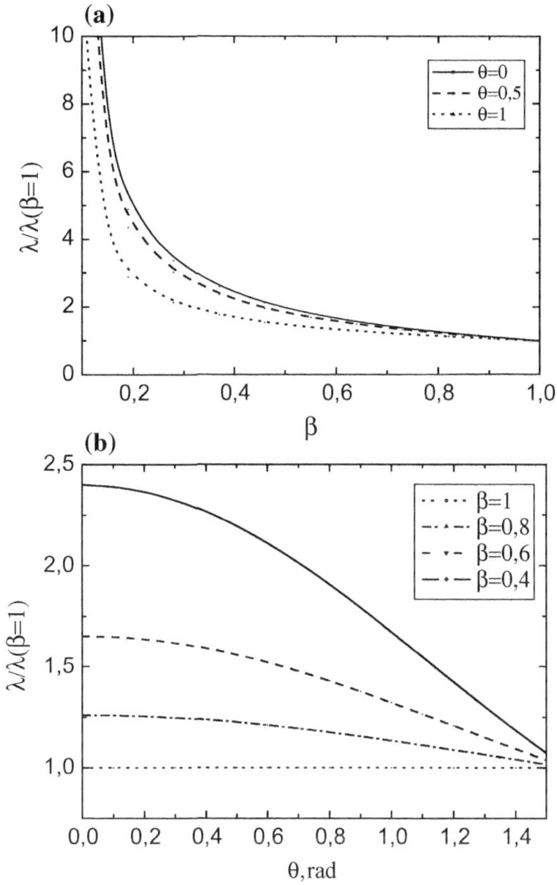

Here it is necessary to clarify what state corresponds to the quantity n_{e0}? Usually, this value corresponds to the equilibrium state, so that n_{e0} does not depend on time, i.e. $dn_{e0}/dt = 0$. However, in the case of a system described by an equation in fractional derivatives, one can not speak of an equilibrium state. Indeed, $\partial^{\beta}_{-\infty\tau} n_{e0} \neq 0$. This means that we are dealing with an unclosed system and we cannot speak of any conserved quantities or stationary processes. Therefore, by n_{e0} we mean the electron concentration at the initial instant of time.

Substituting (4.20) into Eq. (4.19), we obtain:

$$\frac{1}{t_0}\partial^{\beta}_{-\infty\tau}(n_{e0}+\delta n_e) = S - \left(\beta_r(E) + \frac{\partial \beta_r}{\partial E}\delta E\right)(n_{e0}+\delta n_e) - \alpha(n_{e0}+\delta n_e)^2, \quad (4.21)$$

Taking into consideration the fact that

$$\frac{1}{t_0}\partial^{\beta}_{-\infty\tau}n_{e0} = S - \beta_r n_{e0} - \alpha n_{e0}^2, \tag{4.22}$$

we get

$$\frac{1}{t_0}\partial^{\beta}_{-\infty\tau}n_e = -\beta_r \delta n_e - \frac{\partial \beta_r}{\partial E}\delta E n_{e0} - \frac{\partial \beta_r}{\partial E}\delta E \delta n_e - 2\alpha n_{e0}\delta n_e - \alpha \delta n_e^2. \tag{4.23}$$

By linearizing the last equation, we get

$$\frac{1}{t_0}\partial^{\beta}_{-\infty\tau}n_e = -\beta_r \delta n_e - \frac{\partial \beta_r}{\partial E}\delta E n_{e0} - 2\alpha n_{e0}\delta n_e. \tag{4.24}$$

To simplify the equation obtained, we use the following relation, which is valid for small deviations δE and δn_e

$$(n_{e0} + \delta n_e)(E + \delta E) \approx n_{e0}E. \tag{4.25}$$

It follows

$$E\delta n_{e0} = -n_{e0}\delta E. \tag{4.26}$$

We note the general relationship between δE and δn_{e0} can be obtained from physical considerations

$$\delta E = \frac{\delta E}{\delta n_{e0}}\delta n_{e0},$$

where $\delta E/\delta n_{e0} < 0$ that may be explained by dielectric effect—increasing of carrier number leads to decreasing of electric field due to screening effect. In any case, this does not affect on our final result.

Taking (4.26) into account, Eq. (4.24) takes the form

$$\frac{1}{t_0}\partial^{\beta}_{-\infty\tau}(\delta n_e) = E\frac{\partial \beta_r}{\partial E}\delta n_e - \beta_r \delta n_e - 2\alpha n_{e0}\delta n_e. \tag{4.27}$$

We now assume that the time dependence of the perturbation

$$\delta n_e(t) = \delta n_e(0)\mathrm{e}^{-i\omega t}. \tag{4.28}$$

We substitute (4.28) into (4.27). Then we get

$$\omega = i\left[t_0\left(E\frac{\partial \beta_r}{\partial E} - \beta_r - 2\alpha n_{e0}\right)\right]^{1/\beta}. \tag{4.29}$$

The condition for the onset of instability is inequality (4.6). Thus, the problem is reduced to determining the imaginary part of the Eq. (4.29) and investigating this part by condition (4.6).

Here two cases are possible: (1) $E\frac{\partial \beta_r}{\partial E} - \beta_r - 2\alpha n_{e0} > 0$ and (2) $E\frac{\partial \beta_r}{\partial E} - \beta_r - 2\alpha n_{e0} < 0$.

In the first case we have $\operatorname{Im}\omega \geq 0$. Thus, the condition $E\frac{\partial \beta_r}{\partial E} - \beta_r - 2\alpha n_{e0} > 0$ is a condition for the development of instability, but it is not the only. Let us consider the second condition. In this case we rewrite (4.2) in the form

$$\omega = i(-1)^{\beta}\left[t_0\left(\beta_r + 2\alpha n_{e0} - E\frac{\partial \beta_r}{\partial E}\right)\right]^{1/\beta}. \tag{4.30}$$

The expression in square brackets is positive by condition (3.7). From (4.30) we have

$$\operatorname{Im}\omega = \sin\frac{\pi}{2}\left(\frac{2}{\beta}+1\right)\left[t_0\left(\beta_r + 2\alpha n_{e0} - E\frac{\partial \beta_r}{\partial E}\right)\right]^{1/\beta} \geq 0, \tag{4.31}$$

or

$$\sin\frac{\pi}{2}\left(\frac{2}{\beta}+1\right) \geq 0. \tag{4.32}$$

Thus, if condition (4.32) is satisfied, instability develops. If condition (4.32) is not satisfied, then the instability develops only under condition (1). It is interesting that condition (4.32) is a function of a single parameter β, while condition (1) does not depend on this parameter.

The results obtained are of great interest in interpreting real experiments. Within the framework of the existing theory, it is very difficult to take into account many important factors, for example, multiparticle effects or collective phenomena. The latter can be included in the theory by means of a parameter β. In addition, it can be assumed that the process of successive decay of avalanches considered in the work leads to the formation of a fractal structure of the conducting channels. Then, based on the connection between fractal geometry and fractional derivatives [6, 55, 56], it can be said that subsequent processes occurring on this fractal set are described using a fractional derivative, since some of the states in space are excluded for the discharge current to flow.

5 Conclusion

From the discussion above, we can conclude that the appearance of fractional operators in the equation for the Green's function is related with the non-ideality of the system. The consideration was carried in the general form. It can be applied to both fermionic and bosonic systems. The most interesting, in our opinion, is the application of a fractional-differential approach to a phonon gas. As is known, the nonideality of the phonon gas, which is related to the complexity of the crystal structure, leads to anharmonicity, with which many interesting effects are related. The theory of anharmonicity of a fractional character has some peculiarities and can be of great interest in describing real experimental results. For example, the elementary calculation shows that the fractional-differential approach gives a wide class of temperature-dependence of heat capacity. At the same time, it is known that the temperature dependence of the specific heat for complex crystals is not a single-valued function, but is determined by the type of crystal and the structure. The latter circumstance is practically not described by the existing theory. Thus, the fractional-differential approach in the theory of anharmonic effects in crystals is of great interest.

The results obtained are of great interest in interpreting real experiments. Within the framework of the existing theory, it is very difficult to take into account many important factors, for example, many-body effects or collective phenomena. The latter can be included in the theory by means of a parameter. In addition, it can be assumed that the process of successive decay of avalanches considered in the work leads to the formation of a fractal structure of the conducting channels. Then, based on the connection between fractal geometry and fractional derivatives [6, 36, 55, 56], it can be said that subsequent processes occurring on this fractal set are described using a fractional derivative, since some of the states in space are excluded for the discharge current to flow.

Finally, we note that most physical models (classical and quantum) with fractional derivatives are currently at the stage of intensive development. Problems arise even at the stage of choosing one or another fractional operator. In general, preference is given to those models that most adequately describe the available experimental data. The specificity associated with fractional derivatives can manifest itself in the stochastic dynamics and kinetics of large systems. It is logical to assume that simultaneous introduction of fractional derivatives with respect to time and coordinate in the classical and quantum cases is required. However, these issues require further development!

Acknowledgements AZZ, AAM and RGB dedicate this work to the memory of a friend and teacher Prof. Ruslan Meilanov.
AZZ thanks: President grant MK-2130.2017.2, RFBR (# 18-02-01022a).
AAP is grateful to the China grant "Leading Talent Program in Guangdong Province" (No. 00201502, 2016-2020) JiNan University (China, Guangzhou).

References

1. I.M. Lifshitz, M.I. Kaganov, Sov. Phys. Usp. **2**, 831–855 (1960)
2. F.G. Bass, E.A. Rubinstein, Sov. Phys.: Solid State **19**, 800 (1977)
3. F.G. Bass, V.V. Konotop, A.P. Panchekha, JETP 99(6), 1055–1060 (1989)

4. F.G. Bass, A.D. Panchekha, JETP 99(6), 1711–1717 (1991)
5. R. Hilfer (ed.), *Applications of Fractional Calculus in Physics* (World Scientific, 2000), p. 463
6. V.V. Uchaikin, *The method of fractional derivatives* (Publishing house "Artichoke", Ulyanovsk, 2008), p. 512
7. V.V. Uchaikin, Anomalous diffusion and fractional stable distributions. JETP **97**(4), 810–825 (2003)
8. A.A. Stanislavsky, Probability interpretation of the integral of fractional order. Theor. Math. Phys. **138**(3), 418–431 (2004)
9. R.P. Meilanov, M.R. Shabanova, Peculiarities of solutions to the heat conduction equation in fractional derivatives. Tech. Phys. **56**(7), 903–908 (2011)
10. R.T. Sibatov, V.V. Uchaikin, Fractional differential approach to dispersive transport in semiconductors. Phys. Usp. **52**(10), 1019–1043 (2009)
11. SSh Rekhviashvili, Application of fractional integro-differentiation to the calculation of the thermodynamic properties of surfaces. Phys. Solid State **49**(4), 796–799 (2007)
12. X.-B. Wang, J.-X. Li, Q. Jiang, Z.-H. Zhang, Phys. Rev. **49**(14), 9778–9781 (1994)
13. A.V. Milovanov, J.J. Rasmussen, Phys. Rev. B **66**, 134505 (2002)
14. F.J. Dyson, Commun. Math. Phys. **12**(2), 91–107 (1969)
15. F.J. Dyson, Commun. Math. Phys. **12**(3), 212–215 (1969)
16. F.J. Dyson, Commun. Math. Phys. **21**(4), 269–283 (1971)
17. N. Laskin, G. Zaslavsky, Phys. A **368**, 38–54 (2006)
18. V.S. Vladimirov, I.V. Volovich, E.I. Zelenov, *p-adic Analysis and Mathematical Physics* (M.: Nauka, 1994), p. 352 s
19. N. Laskin, Phys. Lett. A **268**, 298–305 (2000)
20. N. Laskin, Phys. Rev. E **62**(3), 3135–3145 (2000)
21. V.M. Eleonsky, V.G. Korolev, N.E. Kulagin, Lett. JETP **76**(12), 859–862 (2002)
22. V.E. Tarasov, Phys. Rev. E **71**(1), 011102 (2005)
23. V.E. Tarasov, Chaos **16**(3), 033108 (2006)
24. V.E. Tarasov, Mod. Phys. Lett. B **21**(5), 237–248 (2007)
25. A.A. Potapov, *Fractals and Scaling in the Radar: A Look from 2015, Book of Abstracts 8nd International Conference (CHAOS' 2015) on Chaotic Modeling, Simulation and Applications*, Henri Poincaré Institute, Paris, 26–29 May 2015, pp. 101, 102
26. A.A. Potapov, V.A. German, Detection of artificial objects with fractal signatures. Pattern Recogn. Image Anal. **8**(2), 226–229 (1998)
27. A.A. Potapov, Y.V. Gulyaev, S.A. Nikitov, A.A. Pakhomov, V.A. German, *Newest Images Processing Methods*, ed. by A.A. Potapov (FIZMATLIT, Moscow, 2008), p. 496
28. Z.Z. Alisultanov, R.P. Meilanov, Theor. Math. Phys. **171**, 404 (2012)
29. Z.Z. Alisultanov, R.P. Meilanov, Theor. Math. Phys. **173**(1), 1445–1456 (2012)
30. Z.Z. Alisultanov, G.B. Ragimkhanov, Fractional-differential approach to the study of instability in a gas discharge. Chaos, Solitons Fractals **107**, 39–42 (2018)
31. A.N. Bogolyubov, A.A. Potapov, S.Sh. Rekhviashvili, Introduction of fractional integro-differentiation in classical electrodynamics. WMU. Phys. Astron. Series 3, 64(4), 9–12 (2009)
32. A.M Nakhushev, *Fractional Calculus and Its Application* (Fizmatlit, Moscow, 2003), p. 272 sec
33. N. Van Hieu, *Fundamentals of the Method of Second Quantization* (Energoatomizdat, Moscow, 1984
34. R. Mailanov, M.Y. Yanpolov, Lett. ZhTF **28**(1), 67–73 (2001)
35. R.P. Meilanov, Z.Z. Alisultanov, *International Russian-Bulgarian Symposium "Equations of the Mixed Type"* (Nalchik, 2010), p. 161
36. A. Potapov, *Fractals in Radiophysics and Radar: Sampling Topology* (The University Book, Moscow, 2005), p. 848
37. M. Naber, *Time fractional Schrodinger equation* Department of Mathematics, Monroe, Michigan, 48161-9746 (2004)
38. L.D. Landau, E.M. Lifshitz, *Statistical Physics, Part 2* (Nauka, Moscow, 1978)
39. G.B. Kadanov, *Quantum Statistical Mechanics* (Mir, Moscow, 1964)

40. C.G. Samko, F.F. Kilbas, O.I. Marichev, *Integrals and Derivatives of Fractional Order and Some Applications* (Science and Technology, Minsk, 1987), p. 688
41. A.B. Alkhasov, R.P. Meilanov, M.R. Shabanova, Fiz **84**(2), 309–317 (2011)
42. I.G. Kaplan, *Introduction to the Theory of Intermolecular Interactions* (Science, Moscow, 1982), p. 312
43. R. Balescu, *Statistical Mechanics of Charged Particles* (Mir, Moscow, 1967), p. 516
44. W. Weibull, J. Appl. Mech.-Trans. ASME. **18**(3), 293–297 (1951)
45. L.D. Landau, E.M. Lifshitz, *Statistical Physics, Part 1* (Nauka, Moscow, 1978)
46. I.A. Kvasnikov, *Thermodynamics and Statistical Physics*, vol. 4. Quantum statistics. - M.: Editorial URSS (2002)
47. A.N. Kozlov, V.N. Flerov, Lett. JETP. **25**(1), 54–58 (1977)
48. V.F. Grin, D.S. Lepsveridze, EA Sal'kov et al., Pis'ma Zh. **87**(8), 415–418 (1975)
49. E.G. Maksimov, Sh Van, M.V. Magnitskaya, S.V. Ebert, Pis'ma Zh. T. **21**(7), 507–510 (2008)
50. Y.E. Lozovik, S.P. Merkulova, A.A. Sokolik, UFN. **178**, 757–776 (2008)
51. G. Savini, A.C. Ferrari, F. Giustino, Phys. Rev. Lett. **105**(037002), 1–4 (2010)
52. A.V. Milovanov, J.J. Rasmussen, arXiv:cond-mat/0201504v3 [cond-mat.supr-con] 8 Oct 2002
53. S.I. Yakovlenko, Tech. Phys. Lett. 2005. T.**31**. AT 4, 76–82
54. V.I. Karelin, A.A. Trenkin, Technical physics. Russ. J. Appl. Phys. **78**(3), 29–35 (2008)
55. A.H. Nielsen, H.L. Pécseli, J. Juul, Phys. Plasmas **3**, 1530 (1996)
56. B.A. Carreras, B. van Milligen, M.A. Pedrosa, R. Balbín, C. Hidalgo, D.E. Newman, E. Sánchez, M. Frances, I. García-Cortés, J. Bleuel, M. Endler, S. Davies, G.F. Matthews, Phys. Rev. Lett. **80**, 4438 (1998)
57. B.A. Carreras, B. van Milligen, C. Hidalgo, R. Balbin, E. Sanchez, I. Garcia-Cortes, M.A. Pedrosa, J. Bleuel, M. Endler, Phys. Rev. Lett. **83**, 3653 (1999)
58. V. Naulin, A. H. Nielsen, J. Juul Rasmussen, Phys. Plasmas **6**, 4575 (1999)
59. G.M. Zaslavsky, M. Edelman, H. Weitzner, B. Carreras, G. McKee, R. Bravenec, R. Fonck, Phys. Plasmas **7**, 3691 (2000)
60. J. Kigami, *Analysis on Fractals* (Cambridge University Press, 2001)
61. R.S. Strichartz, *Differential Equations on Fractals* (Princeton University Press, Princeton and Oxford, 2006)

Similarities Between Dynamics at Atomic and Cosmological Scales

Maricel Agop, Alina Gavriluţ and Gabriel Crumpei

1 On a Multifractal Theory of Motion in a Non-differentiable Space

Since the non-differentiability becomes a fundamental property of the motions space [1–4], a correspondence between the interaction processes and multifractality of the motion trajectories can be established. Then, for all scale resolutions, the geodesics equations (in the form of the Schrödinger equation of fractal type) and some applications (similarities between dynamics at atomic and cosmic scales) are obtained.

2 Consequences of Non-differentiability on a Space Manifold

Let us assume that the motions of the physical systems take place on continuous but non-differentiable curves (fractal curves), so that the following consequences are resulting [5, 6]:

M. Agop
Department of Physics, Gheorghe Asachi Technical University of Iaşi, Iaşi, Romania
e-mail: m.agop@yahoo.com

A. Gavriluţ (✉)
Faculty of Mathematics, Alexandru Ioan Cuza University of Iaşi, Iaşi, Romania
e-mail: gavrilut@uaic.ro

G. Crumpei
Faculty of Psychology and Education Sciences, Alexandru Ioan Cuza University of Iaşi, Iaşi, Romania
e-mail: crumpei.gabriel@yahoo.com

C. H. Skiadas (ed.), *Fractional Dynamics, Anomalous Transport and Plasma Science*, https://doi.org/10.1007/978-3-030-04483-1_8

(i) Any continuous but non-differentiable curve of the physical systems is explicitly scale resolution δt dependent, i.e., its length tends to infinity when δt tends to zero;

We mention that, mathematically speaking, a curve is non-differentiable if it satisfies the Lebesgue theorem [1], i.e. its length becomes infinite when the scale resolution goes to zero. Consequently, in this limit, a curve is as zig-zagged as one can imagine. Thus, it exhibits the property of self-similarity in every one of its points, which can be translated into a property of holography (every part reflects the whole) [1]. This concept of holography can lead to new models for the evolution of cancer, or new models of neural interactions etc.;

(ii) The physics of phenomena is related to the behavior of a set of functions during the zoom operation of the scale resolution δt. Then, through the substitution principle, δt will be identified with dt, i.e., $\delta t \equiv dt$ and, consequently, it will be considered as an independent variable. We reserve the notation dt for the usual time as in the Hamiltonian physical system dynamics.

(iii) The physical system dynamics is described through fractal variables, i.e., functions depending on both the space coordinates and the scale resolution since the differential time reflection invariance of any dynamical variable is broken. Then, in any point of a physical system fractal curve, two derivatives of the variable field $Q(t, dt)$ can be defined:

$$\begin{aligned} \frac{d_+Q(t, dt)}{dt} &= \lim_{\Delta t \to 0_+} \frac{Q(t + \Delta t, \Delta t) - Q(t, \Delta t)}{\Delta t} \\ \frac{d_-Q(t, dt)}{dt} &= \lim_{\Delta t \to 0_-} \frac{Q(t, \Delta t) - Q(t - \Delta t, \Delta t)}{\Delta t} \end{aligned} \tag{1}$$

The "+" sign corresponds to the physical system forward processes, while the "−" sign correspond to the backwards ones;

(iv) The differential of the spatial coordinate field $dX^i(t, dt)$, by means of which we can describe the physical system dynamics, is expressed as the sum of the two differentials, one of them being scale resolution independent (differential part $d_\pm x^i(t)$, and the other one being scale resolution dependent (fractal part $d_\pm \xi^i(t)$), i.e.,

$$d_\pm X^i(t, dt) = d_\pm x^i(t) + d_\pm \xi^i(t, dt); \tag{2}$$

(v) The non-differentiable part of the spatial coordinate field, by means of which we can describe the physical system dynamics, satisfies the fractal equation:

$$d_\pm \xi^i(t, dt) = \lambda^i_\pm (dt)^{1/D_F} \tag{3}$$

where $\lambda^i_\pm$ are constant coefficients through which the fractalization type describing the physical system dynamics is specified and D_F defines the fractal dimension of the physical system non-differentiable curve.

In our opinion, physical systems processes imply dynamics on geodesics with various fractal dimensions. The variety of these fractal dimensions of the physical systems geodesics comes as a result of its structure. Precisely, for $D_F = 2$, quantum type processes are generated in a physical system. For $D_F < 2$ correlative type processes are induced, while for $D_F > 2$ non-correlative type ones can be found - for details see [5].

Because all the processes described here can take place simultaneously in the dynamics of a physical system, it is thus necessary to consider the multi-fractal behavior of physical structures (for details see [2–4]);

(vi) The differential time reflection invariance of any dynamical variable of the physical system is recovered by combining the derivatives d_+/dt and d_-/dt in the non-differentiable operator:

$$\frac{\hat{d}}{dt} = \frac{1}{2}(\frac{d_+ + d_-}{dt}) - \frac{i}{2}(\frac{d_+ - d_-}{dt}) \tag{4}$$

This is a natural result of the complex prolongation procedure applied to physical system dynamics [7]. Applying now the non-differentiable operator to the spatial coordinate field, by means of which we can describe the physical system dynamics, yields the complex velocity field:

$$\hat{V}^i = \frac{\hat{d}X^i}{dt} = V_D^i - V_F^i \tag{5}$$

with

$$V_D^i = \frac{1}{2}\frac{d_+X^i + d_-X^i}{dt}, \; V_F^i = \frac{1}{2}\frac{d_+X^i - d_-X^i}{dt} \tag{6}$$

The real part V_D^i of the complex velocity field is differentiable and scale resolution independent (differentiable velocity field), while the imaginary one V_F^i is non-differentiable and scale resolution dependent (fractal velocity field).

(vii) In the absence of any external constraint, an infinite number of fractal curves (geodesics) can be found relating any pair of points, and this is true on all scales of the physical system dynamics. Then, in the fractal space of the physical system, all its entities are substituted with the geodesics themselves so that any external constraint can be interpreted as a selection of geodesics. The infinity of geodesics in the bundle, their non-differentiability and the two values of the derivative imply a generalized statistical fluid-like description (in what follows we shall call it a fractal fluid). Then, the average values of the biological fractal fluid variables must be considered in the previously mentioned sense, so the average of $d_\pm X^i$ is:

$$< d_\pm X^i > \equiv d_\pm x^i \tag{7}$$

with

$$< d_\pm \xi^i > = 0 \tag{8}$$

The previous relation (8) implies that the average of the fractal fluctuations is null.

(viii) The fractal fluid dynamics can be described through a scale covariant derivative, the explicit form of which is obtained as follows. Let us consider that the non-differentiable curves are immersed in a 3 -dimensional space and that X^i are the spatial coordinate field of a point on the non-differentiable curve. We also consider a variable field $Q(X^i, t)$ and the following Taylor expansion up to the second order:

$$d_{\pm}Q(X^i, t) = \partial_t Q dt + \partial_i Q d_{\pm}X^i + \frac{1}{2}\partial_l \partial_k Q d_{\pm}X^l d_{\pm}X^k \tag{9}$$

These relations are valid in any point and more for the points X^i on the non-differentiable curve which we have selected in (9). From here, the main forward and backward values for fractal fluid variables from (9) become:

$$< d_{\pm}Q >=< \partial_t Q dt > + < \partial_i Q d_{\pm}X^i > +\frac{1}{2} < \partial_l \partial_k Q d_{\pm}X^l d_{\pm}X^k > \tag{10}$$

We suppose that the average values of the all variable field Q and its derivatives coincide with themselves and the differentials $d_{\pm}X^i$ and dt are independent. Therefore, the average of their products coincides with the product of averages. Consequently, (10) becomes:

$$d_{\pm}Q = \partial_t Q dt + \partial_i Q < d_{\pm}X^i > +\frac{1}{2}\partial_l \partial_k Q < d_{\pm}X^l d_{\pm}X^k > \tag{11}$$

Even the average value of $d_{\pm}\xi^i$ is null, for the higher order of $d_{\pm}\xi^i$ the situation can still be different. Let us focus on the averages $< d_{\pm}\xi^l d_{\pm}\xi^k >$. Using (3) we can write:

$$< d_{\pm}\xi^l d_{\pm}\xi^k > \pm\lambda^l_{\pm}\lambda^k_{\pm}(dt)^{(2/D_F)-1}dt \tag{12}$$

where we accepted that the sign $+$ corresponds to $dt > 0$ and the sign $-$ corresponds to $dt < 0$.

Then, (11) takes the form:

$$\begin{aligned} d_{\pm}Q = \partial_t Q dt + \partial_i Q < d_{\pm}X^i > +\frac{1}{2}\partial_l \partial_k Q d_{\pm}x^l d_{\pm}x^k \\ \pm \frac{1}{2}\partial_l \partial_k Q[\lambda^l_{\pm}\lambda^k_{\pm}(dt)^{(2/D_F)-1}dt] \end{aligned} \tag{13}$$

If we divide by dt and neglect the terms that contain differential factors (for details, see the method from [6]) we obtain:

$$\frac{d_{\pm}Q}{dt} = \partial_t Q + v^i_{\pm}\partial_i Q \pm \frac{1}{2}\lambda^l_{\pm}\lambda^k_{\pm}(dt)^{(2/D_F)-1}\partial_l \partial_k Q \tag{14}$$

These relations also allow us to define the operators

$$\frac{d_{\pm}}{dt} = \partial_t + v_{\pm}^i \partial_i \pm \frac{1}{2} \lambda_{\pm}^l \lambda_{\pm}^k (dt)^{(2/D_F)-1} \partial_l \partial_k \tag{15}$$

where

$$v_+^i = \frac{d_+ x^i}{dt}, \; v_-^i = \frac{d_- x^i}{dt}$$

Under these circumstances, taking into account (4), (5) and (15), let us calculate $\hat{d}/dt$. It results:

$$\frac{\hat{d}Q}{dt} = \partial_t Q + \hat{V}^i \partial_i Q + \frac{1}{4}(dt)^{(2/D_F)-1} D^{lk} \partial_l \partial_k Q \tag{16}$$

where

$$\begin{aligned} D^{lk} &= d^{lk} - i\overline{d}^{lk} \\ d^{lk} &= \lambda_+^l \lambda_+^k - \lambda_-^l \lambda_-^k, \overline{d}^{lk} = \lambda_+^l \lambda_+^k + \lambda_-^l \lambda_-^k \end{aligned} \tag{17}$$

The relation (16) also allows us to define the scale covariant derivative in the fractal fluid dynamics

$$\frac{\hat{d}}{dt} = \partial_t + \hat{V}^i \partial_i + \frac{1}{4}(dt)^{(2/D_F)-1} D^{lk} \partial_l \partial_k \tag{18}$$

3 Fractal Fluid Geodesics

Let us now consider the principle of scale covariance (the physics laws, which are specific to the physical fractal fluid dynamics, are invariant with respect to scale transformations) and postulate that the passage from the classical (differentiable) physics to the fractal (non-differentiable) physics can be implemented by replacing the standard time derivative d/dt with the non-differentiable operator $\hat{d}/dt$. Thus, this operator plays the role of the scale covariant derivative, namely it is used to write the fundamental equations of fractal fluid dynamics in the same form as in the classic (differentiable) case. Under these conditions, applying the operator (18) to the complex velocity field (5), in the absence of any external constraint, the geodesics take the following form:

$$\frac{\hat{d}\hat{V}^i}{dt} = \partial_t \hat{V}^i + \hat{V}^l \partial_l \hat{V}^i + \frac{1}{4}(dt)^{(2/D_F)-1} D^{lk} \partial_l \partial_k \hat{V}^i = 0 \tag{19}$$

This means that the local acceleration $\partial_t \hat{V}^i$, the local convection $\hat{V}^l \partial_i \hat{V}^i$ and the local dissipation $D^{lk} \partial_l \partial_k \hat{V}^i$, make their balance in any point of the non-differentiable curve. Moreover, the presence of the complex coefficient of viscosity-type in

$\frac{1}{4}(dt)^{(2/D_F)-1}D^{lk}$ in the fractal fluid dynamics specifies that it is a rheological medium. So, it has memory, as a datum, by its own structure.

If the fractalisation is achieved by Markov type stochastic processes, which involve Lévy type movements [1–4] of the fractal fluid entities, then:

$$\lambda_+^i \lambda_+^l = \lambda_-^i \lambda_-^l = 2\lambda\delta^{il} \tag{20}$$

where δ^{il} is the Kronecker's pseudo-tensor.

Under these conditions, the geodesic equation of the fractal fluid takes the simple form

$$\frac{\hat{d}\hat{V}^i}{dt} = \partial_t \hat{V}^i + \hat{V}^l \partial_l \hat{V}^i - i\lambda(dt)^{(2/D_F)-1}\partial^l \partial_l \hat{V}^i = 0 \tag{21}$$

or more, by separating the motions on differential and fractal scale resolutions,

$$\begin{aligned}
\frac{\hat{d}V_D^i}{dt} &= \partial_t V_D^i + V_D^l \partial_l V_D^i - [V_F^l + \lambda(dt)^{(2/D_F)-1}\partial^l]\partial_l V_F^i = 0 \\
\frac{\hat{d}V_F^i}{dt} &= \partial_t V_F^i + V_D^l \partial_l V_F^i + [V_F^l + \lambda(dt)^{(2/D_F)-1}\partial^l]\partial_l V_D^i = 0
\end{aligned} \tag{22}$$

4 Fractality and Its Implications

The separation of the physical system dynamics on scale resolutions specifies at the differentiable scale resolutions the fractal force:

$$F_F^i = (V_F^l + \lambda(dt)^{(2/D_F)-1}\partial^l)\partial_l V_F^i \tag{23}$$

Its cancellation

$$(V_F^l + \lambda(dt)^{(2/D_F)-1}\partial^l)\partial_l V_F^i = 0 \tag{24}$$

on the condition

$$\partial_l V_F^l = 0 \tag{25}$$

induces a particular velocities field whose explicit form will be given in what follows.

Finding the solutions for these equations can be relatively difficult, due to the fact that this equation system is a non-linear one. However, there is an analytical solution of this system, in the particular case of a "stationary flow" in a plane symmetry (x, y). In these circumstances, Eqs. (24) and (25) take the form:

$$V_x \frac{\partial V_x}{\partial x} + V_y \frac{\partial V_x}{\partial x} = \lambda(dt)^{(2/D_F)-1}\frac{\partial^2 V_x}{\partial y^2} \tag{26}$$

$$\frac{\partial V_x}{\partial x} + \frac{\partial V_y}{\partial y} = 0 \tag{27}$$

where $V_{Fx} = V_x(x, y)$ is the velocity along axis Ox, $V_{Fy} = V_y(x, y)$ is the velocity along axis Oy. The boundary condition of the flow are:

$$\lim_{y \to 0} V_y(x, y) = 0, \lim_{y \to 0} \frac{\partial V_x}{\partial y} = 0, \lim_{y \to \infty} V_x(x, y) = 0 \tag{28}$$

and the flux momentum per length unit is constant:

$$\Theta = \rho \int_{-\infty}^{+\infty} V_x^2 dy = \text{const.} \tag{29}$$

Using the method from [8] for solving the Eqs. (26) and (27), with the limit conditions (28) and (29), the following solutions result:

$$V_x = \frac{[1,5(\frac{\Theta}{6\rho})^{2/3}]}{[\lambda(dt)^{(2/D_F)-1}x]^{1/3}} \cdot \sec\text{h}^2 \frac{[(0,5y)(\frac{\Theta}{6\rho})^{1/3}]}{[\lambda(dt)^{(2/D_F)-1}x]^{2/3}} \tag{30}$$

$$V_y = \frac{[4,5(\frac{\Theta}{6\rho})^{2/3}]}{[3\lambda(dt)^{(2/D_F)-1}x]^{1/3}} \cdot [\frac{y(\frac{\Theta}{6\rho})^{1/3}}{[\lambda(dt)^{(2/D_F)-1}x]^{2/3}} \cdot \\ \cdot \sec\text{h}^2 \frac{[(0,5y)(\frac{\Theta}{6\rho})^{1/3}]}{[\lambda(dt)^{(2/D_F)-1}x]^{2/3}} - \tanh \frac{[(0,5y)(\frac{\Theta}{6\rho})^{1/3}]}{[\lambda(dt)^{(2/D_F)-1}x]^{2/3}}] \tag{31}$$

Relations (30) and (31) suggest that at all scale resolutions, the fractal fluid velocity field is highly non-linear by means of soliton and soliton-kink type solutions (for details see [9]).

For $y = 0$, we obtain in relation (30) the flow critical velocity in the form:

$$V_x(x, y = 0) = V_c = \frac{[1,5(\frac{\Theta}{6\rho})^{2/3}]}{[\lambda(dt)^{(2/D_F)-1}x]^{1/3}} \tag{32}$$

while relation (29), taking into account (32) becomes:

$$\Theta = \rho \int_{-\infty}^{+\infty} V_x^2(x, y) dy = \int_{-d_c}^{+d_c} V_c^2(x, 0) dy \tag{33}$$

so that the critical cross section of the strains lines tube is given by:

$$d_c(x, y = 0) = \frac{\Theta}{2\rho V_c^2} = 2,42[\lambda(dt)^{(2/D_F)-1}x]^{2/3}(\frac{\rho}{\Theta})^{1/3} \tag{34}$$

Relations (30) and (31) can be strongly simplified if we introduce the normalized quantities:

$$u = V_x \frac{y_0^2}{x_0 \lambda}, v = V_y \frac{y_0^2}{x_0 \lambda}, \xi = \frac{x}{x_0}, \eta = \frac{y}{y_0}, \mu = \lambda (dt)^{(2/D_F)-1}, \frac{(\frac{\Theta}{6\rho})^{1/3}}{\lambda^{2/3}} = \frac{x_0^{2/3}}{y_0} \tag{35}$$

where x_0, y_0 are specific lengths of the fractal fluid flow. It results:

$$u(\xi, \eta) = \frac{1,5}{\xi^{1/3}\mu^{1/3}} \cdot \sec\mathrm{h}^2(\frac{0,5}{\mu^{2/3}} \frac{\eta}{\xi^{2/3}}) \tag{36}$$

$$v(\xi, \eta) = \frac{(4,5)^{2/3}}{3^{1/3}\mu^{1/3}\xi^{1/3}}[\frac{\eta}{\mu^{2/3}\xi^{2/3}} \sec\mathrm{h}^2(\frac{0,5}{\mu^{2/3}} \frac{\eta}{\xi^{2/3}}) - \tanh(\frac{0,5}{\mu^{2/3}} \frac{\eta}{\xi^{2/3}})] \tag{37}$$

In Fig. 1 we present the dependencies on the dimensionless spatial coordinates ξ, η both of the fractal velocities components $u = u(\xi, \eta)$, $v = v(\xi, \eta)$ and of the fractal velocity modulus $V = \sqrt{u^2 + v^2} = V(\xi, \eta)$ for the fractality degrees $\mu = 0.5; 0.9; 4$. We observe an expansion of the fractal velocity field, both on the components and on its modulus together with the increase of the fractality degree.

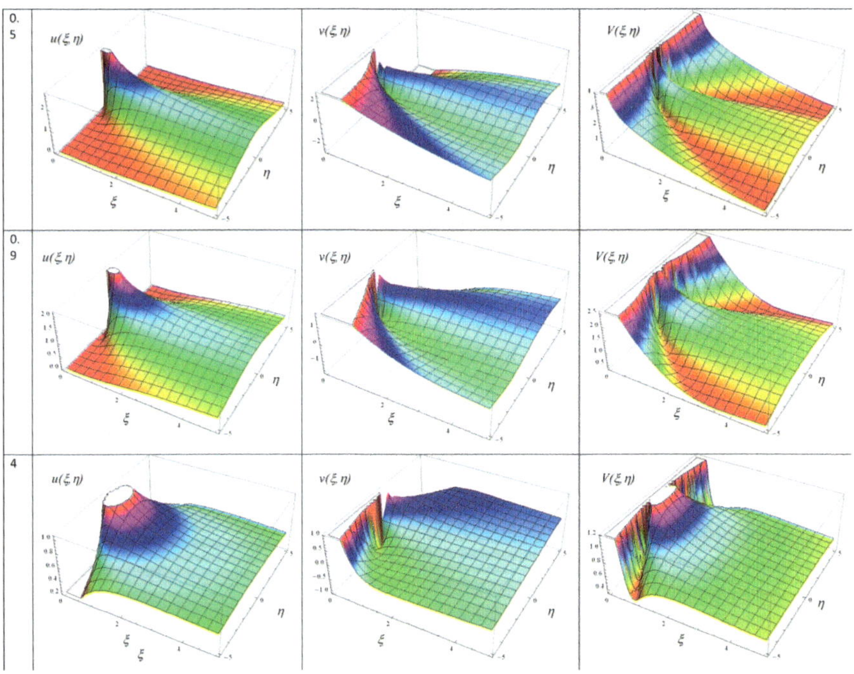

Fig. 1 The three-dimensional dependencies $u = u(\xi, \eta)$, $v = v(\xi, \eta)$, $V = \sqrt{u^2 + v^2} = V(\xi, \eta)$ for the fractality degrees $\mu = 0.5; 0.9; 4$.

5 Fractal Geodesics in the Schrödinger Type Representation. Applications

Fractal Geodesics

For irrotational motions of the fractal fluid, the complex velocity field $\hat{V}^l$ takes the form:

$$\hat{V}^l = -2i\lambda(dt)^{(2/D_F)-1}\partial^l \ln \Psi \tag{38}$$

Then the geodesics Eq. (21) becomes:

$$\begin{aligned}\frac{\hat{d}\hat{V}^l}{dt} &= -2i\lambda(dt)^{(2/D_F)-1}\partial_t\partial^l \ln \Psi \\ &+ [-2i\lambda(dt)^{(2/D_F)-1}\partial_p \ln \Psi \\ &- i\lambda(dt)^{(2/D_F)-1}\partial_p]\partial^p\partial^l[-2i\lambda(dt)^{(2/D_F)-1} \ln \Psi] = 0.\end{aligned} \tag{39}$$

Since

$$\begin{aligned}&\partial^l(\partial_p \ln \Psi \partial^p \ln \Psi) = 2\partial_p \ln \Psi \partial^p \partial^l \ln \Psi \\ &\partial^l \partial_p \partial^p \ln \Psi = \partial_p \partial^p \partial^l \ln \Psi \\ &\partial^l(\partial_p \ln \Psi \partial^p \ln \Psi) + \partial_p \partial^p \ln \Psi) = \partial^l(\frac{\partial_p \partial^p \Psi}{\Psi})\end{aligned} \tag{40}$$

Eq. (39) becomes:

$$i\lambda(dt)^{(2/D_F)-1}\partial_t\partial^l \ln \Psi + \lambda^2(dt)^{(4/D_F)-2}\partial^l(\frac{\partial_p\partial^p\Psi}{\Psi}) = 0 \tag{41}$$

By integrating the above relation we obtain:

$$\lambda^2(dt)^{(4/D_F)-2}\partial_p\partial^p\Psi + i\lambda(dt)^{(2/D_F)-1}\partial_t\Psi + Q^2(t)\Psi = 0 \tag{42}$$

with $Q^2(t)$ an arbitrary function of t.

In consequence, the non-differentiable geodesics (42) in the terms of Ψ are defined up to an arbitrary function $Q^2(t)$, which is dependent on t . For $Q(t) = 0$, the relation (42) reduces to the fractal type Schrödinger equation:

$$\lambda^2(dt)^{(4/D_F)-2}\partial_p\partial^p\Psi + i\lambda(dt)^{(2/D_F)-1}\partial_t\Psi = 0 \tag{43}$$

The standard Schrödinger equation:

$$\frac{\hbar^2}{2m_0}\partial_p\partial^p\Psi + i\hbar\partial_t\Psi = 0$$

can be obtained from (43) for non-relativistic motions on Peano curves [1], $D_F = 2$, at Compton scale $\lambda = \hbar/2m_0$, with $\hbar$ the Planck reduced constant and m_0 the rest mass of the physical entity.

In the case of non-differentiable dynamics with constraints, for instance under the action of a scalar potential U, following the same procedure as before, one obtains the fractal type equation:

$$\lambda^2 (dt)^{(4/D_F)-2} \partial^l \partial_l \Psi + i\lambda (dt)^{(2/D_F)-1} \partial_t \Psi - \frac{U}{2}\Psi = 0 \tag{44}$$

For non-relativistic motions on Peano curves, $D_F = 2$, at Compton scale $\lambda = \hbar/2m_0$, the Eq. (44) takes the standard form:

$$\frac{\hbar^2}{2m_0}\partial_l \partial^l \Psi + i\hbar \partial_t \Psi - U\Psi = 0$$

If the scalar potential U is time independent, the Eq. (44) admits the stationary solution:

$$\Psi(\mathbf{r}, t) = \Psi(\mathbf{r}) \exp[-\frac{i}{2m_0\lambda (dt)^{(2/D_F)-1}} Et] \tag{45}$$

where E is the fractal energy of the physical system of fractal type. Then $\Psi(\mathbf{r})$ is the solution of non-temporal Schrödinger equation of fractal type:

$$\partial_l \partial^l \Psi + \frac{1}{2m_0\lambda^2 (dt)^{(4/D_F)-2}}(E - U)\Psi = 0 \tag{46}$$

6 Fractal Motions in Central Field

Let us consider the case of a field of the form (central field)

$$U(r) = -\frac{C}{r}, C = \text{const.} \tag{47}$$

Since this field has spherical symmetry, the fractal type Schrödinger stationary Eq. (46) in the presence of the scalar potential (47) becomes:

$$\begin{aligned} &\frac{\partial}{\partial r}(r^2 \frac{\partial \Psi}{\partial r}) + \frac{1}{2m_0\lambda^2 (dt)^{(4/D_F)-2}}(E + \frac{C}{r})\Psi \\ &= -[\frac{1}{\sin\theta}\frac{\partial}{\partial\theta}(\sin\theta \frac{\partial\Psi}{\partial\theta}) + \frac{1}{\sin^2\theta}\frac{\partial^2\Psi}{\partial\varphi^2}] \end{aligned} \tag{48}$$

Following a procedure similar to the one from [10], one obtains the eigen-solution:

$$\Psi_{nlm}(r,\theta,\varphi) = R_{nl}(r)Y_{ml}(\theta,\varphi) \tag{49}$$

with

$$\begin{aligned} R_{nl}(r) &= \left(\frac{2}{nr_0}\right)^{3/2}\left\{\frac{(n-l-1)!}{2n[(n+l)!]^3}\right\}^{1/2}\left(\frac{2r}{nr_0}\right)^l \exp\left(-\frac{r}{nr_0}\right)L_{n+l}^{2l+1}\left(\frac{2r}{nr_0}\right) \\ Y_{ml}(\theta,\varphi) &= (-1)^k\left[\frac{(l-m)!(2l+1)}{(l+|m|)!4\pi}\right]^{1/2} P_l^m(\cos\theta)\exp(im\varphi) \\ r_0 &= \frac{4m_0}{C}\lambda^2(dt)^{(4/D_F)-2} \end{aligned} \tag{50}$$

$$n = 1,2,3,\ldots, l = 0,1,2,\ldots,n-1, m = 0,\pm 1,\pm 2,\ldots,\pm l$$

$$k = \begin{cases} m, & \text{for } m \geq 0 \\ 0, & \text{for } m < 0, \end{cases}$$

where L_{n+l}^{2l+1} are the generalized Laguerre polynomials, P_l^m are the generalized Legendre polynomials, n is the principal fractal number, l is the orbital fractal number, m is the magnetic fractal number, respectively the eigenvalue:

$$E_n = -2m_0\lambda^2(dt)^{(4/D_F)-2}\frac{1}{n^2r_0^2} = -\frac{1}{n^2}\cdot\frac{C^2}{8m_0\lambda^2(dt)^{(4/D_F)-2}} \tag{51}$$

Now, if we rewrite the Eq. (51) in one of the forms:

$$E_n = -\frac{m_0}{2}(v_n)^2 = -\frac{m_0}{2}\left(\frac{v_0}{n}\right)^2 \tag{52}$$

one obtains the fractal law of orbital velocities quantification:

$$v_n = \frac{v_0}{n}, v_0 = \frac{2\lambda(dt)^{(2/D_F)-1}}{r_0} = \frac{C}{2m_0\lambda(dt)^{(2/D_F)-1}} \tag{53}$$

Moreover, accepting the functionality of the relation:

$$\frac{m_0v_n^2}{r_n} = \frac{C}{r_n^2} \tag{54}$$

one finds the fractal law of the orbital rays quantification in the form:

$$r_n = n^2r_0, r_0 = \frac{C}{m_0v_0^2} \tag{55}$$

Based on the relations (53) and (55) one obtains both the fractal law of quantification of the orbital specific kinetic moments:

$$l_n = r_n v_n = n l_0,\, l_0 = 2\lambda (dt)^{(2/D_F)-1} \tag{56}$$

and the fractal law of quantification of the orbital periods:

$$T_n = 2\pi \frac{r_n}{v_n} = n^3 T_0,\, T_0 = 2\pi \frac{r_0}{v_0} = 16\pi m_0^2 \frac{\lambda^3 (dt)^{(6/D_F)-3}}{C^2} \tag{57}$$

Certain similarities between the dynamics at small scale resolution (atomic scale) and high scale resolution (cosmological scale) are presented in Table 1.

In Table 1, e is the fundamental electric charge, ε is the vacuum electric permittivity, G is the gravitational constant, M is the mass of the source body, m is the test particle mass and w is the velocity "quanta" in the gravitational structures [11].

7 Quantifiable Dynamics at Infragalactic Scale Resolutions. Theoretical and Experimental Aspects

The quantifiable dynamics at infragalactic scale resolution refer to motions of the gravitational structures inside the galaxies (for instance, the planetary motions around the Sun, the motions of the natural satellites around the planets, the motions of the stars inside the clusters of galaxies etc.). In what follows, taking into account the astronomical observations, we shall prove that at our solar system level, the dynamics (both of the planets around the Sun and of the natural satellites around the planets) are quantifiable. Indeed, in Table 2 we show the quantification of planetary motion using the relations (53), (55), (56) and (57) comparing with the experimental data. Moreover, the quantification of the Saturn's rings motion is also given - see Table 3.

It should be noted that Eqs. (53) and (56) and their consequences only determine the allowed orbits. Actually, only a few allowed orbits are occupied by the planets. The reason for this can only be subjected to guesses (in fact, the origin of the planets is yet today a subject of guesses). One can suppose for instance that the protoplanetary material was originally distributed to form rings in correspondence with every fractal number, and that gravitational instabilities or perturbations have brought to a final state of aggregation of the material in planets over some orbits in correspondence with some particular fractal numbers. Otherwise, one can guess that external gravitational perturbations acting on an almost continuum protoplanetary material in gravitational, centrifugal and thermodynamical equilibrium, led to spatial density resonances mainly corresponding to some transversal eigenfunctions (which, of course, must satisfy circular periodicity conditions).

Anyway, and with a certain surprise one can see from Table 2 that the interpretation of the planetary orbits with a gravitational atom model is in reasonable

Table 1 Similarities between dynamics at atomic scale (of electromagnetic type) and at cosmological scale (of gravitational type)

	Small scale resolution(atomic scales)		High scale resolution(cosmological scales)	
	Constants of structure	Fractal laws of quantification	Constants of structure	Fractal laws of quantification
Constants of interaction	$C_l = \frac{e^2}{4\pi\varepsilon_0}$		$C_G = GM$	
Coefficients associated to the transition fractal-non-fractal	$\lambda_l = \frac{h}{4\pi m}$		$\lambda_G = \frac{GM}{w}$, $w = (144, 7 \pm 0, 3)\text{km/s}$	
Fractal laws of quantification of the orbital velocities	$v_{0e} = \frac{e^2}{2\varepsilon_0 h}$	$v_{en} = \frac{v_{0e}}{n}$	W	$v_{Gn} = \frac{w}{n}$
Fractal laws of quantification of the orbital rays	$r_{oe} = \frac{\varepsilon_0 h^2}{\pi m e^2}$	$r_{er} = n^2 v_{0e}$	$r_{0G} = \frac{GM}{w}$	$r_{Gn} = n^2 V_{0G}$
Fractal laws of quantification of the orbital specific kinetic momentum	$l_{0e} = \frac{h}{2\pi m}$	$l_{en} = n l_{0e}$	$l_{0G} = \frac{GM}{w}$	$l_{Gn} = n l_{0G}$
Fractal laws of quantification of the orbital periods	$T_{0e} = \frac{2\varepsilon_0^2 h^3}{\pi m_0 e^4}$	$T_{en} = n^3 T_{0e}$	$T_{0G} = \frac{GM}{w^3}$	$T_{Gn} = n^3 T_{0G}$
Fractal laws of quantification of the orbital energies	$E_{0l} = \frac{m_0 e^4}{8\varepsilon^2 h^2}$	$E_{en} = -\frac{E_{0e}}{n^2}$	$E_{0G} = m_0 \frac{w^2}{2}$	$E_G = -\frac{E_{0G}}{n^2}$

Table 2 Quantization of the planetary system

Planet:		–	Mercury	Venus	Earth	Mars	–	Ceres	–	Jupiter	–	Saturn	–	Uranus	–	Neptune	–	Pluto
m		1–2	3	4	5	6	7	8	9–10	11	12–14	15	16–20	21	22–25	26	27–29	30
$r_m v_m \times 10^{15} (m^2 s^{-1})$	Theor.	–	2.75	3.67	4.58	5.50	–	7.34	–	10.09	–	13.76	–	19.27	–	23.86	–	27.53
	Exp.	–	2.76	3.78	4.45	5.49	–	–	–	10.15	–	13.75	–	19.55	–	24.55	–	27.91
$v_m (\mathrm{kms}^{-1})$	Theor.	–	47.9	35.9	28.7	23.9	–	17.9	–	13.0	–	9.5	–	6.8	–	5.5	–	4.7
	Exp.	–	47,9	35.0	29.8	24.1	–	–	–	13.0	–	9.0	–	6.8	–	5.4	–	4.7
r_m(a.u.)	Theor.	–	0.383	0.681	1.065	1.533	–	2.726	–	5.154	–	9.585	–	18.786	–	28.797	–	38.340
	Exp.	–	0.387	0.723	1.0	1.523	–	2.766	–	5.202	–	9.536	–	19.210	–	30.138	–	39.390
T_m(year)	Theor.	–	0.238	0.565	1.103	1.907	–	4.52	–	11.752	–	29.801	–	81.774	–	155.196	–	238.41
	Exp.	–	0.241	0.615	1	1.881	–	4.67	–	11.962	–	29.458	–	84.013	–	164.794	–	247.686

$r_0 v_0 = 0.9178 \times 10^{15}\ \mathrm{m}^2\ \mathrm{s}^{-1}$; $w = 144.7\,\mathrm{km\,s}^{-1}$; $r_0 = 0.0426$ a.u.; $T_0 = 0.00883$ year

Table 3 Quantization of the orbital radii of Saturn's rings

Rings:		–	D_{IE}	C_{IE}	B_{IE}	Cassini	A_{IE}	F	G	E_{IE}	–	E_{OE}
m		1–5	6	–	7	–	8	9	–	10	11–15	16
$r_m \times 10^3$ (km)	Theor.	–	65.4	–	89.0	–	116.2	147.1	–	181.7	–	465.1
	Exp.	–	67.0	74.5	92.0	119.8	122.2	140.4	170	180.0	–	480.0

$r_0 = 1.817 \times 10^3$ km

agreement with the observed data, even for small fractal principal numbers. However, we acknowledge that it is still not enough to give validity to our conjecture, and that alternative hypotheses could be formulated. To extend the investigation on this topic, some more information about other planetary systems is required" [11].

In [12] the planetary motion of a companion of the star 51 Pegasus, with a mass ranging between one half and twice the mass of Jupiter and mean distance from 51 Pegasus about 0.05 a.u., has been examined: it turns out that its orbit nearly corresponds to the fractal number $n = 1$ of its planetary system. Some reported uncertainty on the mass of 51 Peg (a star "quite similar to the Sun") could be removed if one guesses that the orbit of its large planet exactly corresponds to $n = 1$: on the basis of the observed period ($P = 4.23$ days) a distance $r_0 = 0.055$ a.u. and a mass $M^{51\ \mathrm{Peg.}} = 1.26M_\odot$ can be obtained. Thus, some recent knowledge about a simple "gravitational atom" different from our solar system seems to support our guess. In particular the orbits of the planets $HR5185$ and $HR458$ corresponds to $n = 1$.

We will shortly examine some consequences of our guess on the allowable orbits around the various planets. Of course, the major semi-axes of these orbits are obtained by substituting in Eq. (55) the mass of the Sun $M_\odot$ with the mass M_P of the concerned planet. Assuming the universality of the gravitational structure constant $\alpha = c/w$, with c the velocity of light in the vacuum, one immediately gets from Eq. (55)

$$r_0^{(P)} = r_0^\odot M_P/M_\odot$$

where $r_0^{(P)}$ indicates the "Bohr radius" for orbiting motion around the planet.

Using the value $r_0^\odot = 0.0439 \pm 0.0004$ a.u. (for details see [11]) one can obtain some interesting results. For instance, considering the principal fractal number of the first "free" (i.e., out of the planet) allowable orbits, one can note that for Mercury, Venus, the Earth, Mars and Pluto these orbits correspond to relatively high (≥ 19) fractal principal numbers. The opposite, i.e., small (≤ 11) "free" fractal numbers, is verified for the big planets.

As regards the orbits of the satellites one really obtains both good and bad results. This disappointing ambiguity could be partly attributed to the important role that other forces, besides the gravitational one, play in this case.

More interesting coincidences are found if one considers the Saturn ring "powder" structure. In Table 2 we show the main features of this structure and the corresponding discretized radii. One can note that the inner edges of the main rings (D, B and the Cassini division together with the neighboring A ring) are almost exactly singled out by the corresponding calculated radii. An exception is represented by the inner edge of the C ring. On the other hand, this ring does not originate from purely gravitational forces, since, as explicitly remarked in the Landolt-Börnstein handbook [13], the boundary between the B and C rings is presumably related to electrodynamical interaction between the charged dust particles of the ring system and the planetary magnetic field. One must finally note the remarkable calculated singling out of the (somewhat conventional) limits of the broad rarefied E ring.

Of course, we agree on the fact that the previous remarks about the orbital motion around the planets are in no way conclusive in order to check our guess, and in particular the "universality" of the gravitational structure constant α_g, which, we recall, has been estimated by calculations concerning only orbital motions around the Sun [11].

8 Quantifiable Dynamics at Extragalactic Scale Resolutions. Theoretical and Experimental Aspects

The dynamics at extragalactic scale resolutions refer to intergalactic motions.

The application of our fractal method to the problem of galaxy pairs is similar to the solar system case. We start from the remark that even an "isolated" pair in galaxy catalogues is never truly isolated. We shall then describe the effect of the uncontrollable interactions of the environment in terms of the here above complex Wiener process, i.e., for $D_F = 2$.

In classical as well as quantum mechanics and more generally as fractal mechanics, the problem of the relative motion of two bodies can be reduced in the reference system of the center of inertia to that of one body of mass,

$$\mu = m_1 m_2/(m_1 + m_2)$$

In the particular case of gravitation, the potential $\Phi_G = -Gm_1m_2/r$, so that the equation of motion is that of a test particle around a body of mass $m_0 = m_1 + m_2$. The same is true here, so that Eq. (48) still applies to this case.

We find that the pair energy is quantized as:

$$E_{G_n} = -\frac{1}{2}\mu\frac{w^2}{2}$$

and that the relative velocity in binary galaxies must take only preferential values given by:

$$v_{G_n} = \frac{w}{n},\ w = \alpha_g \cdot c,\ \alpha_g^{-1} = 2072 \pm 7$$

We enumerate in Table 4 the velocity quantization for some typical pairs of galaxies. Such a theoretical result seems to provide an explanation for Tifft's effect of red-shift quantization in binary galaxies [14–16]. Indeed it has been claimed by Tifft [16] that the speed differences in isolated galaxy pairs was not distributed at random, but showed preferential values near 144, 72, 36 and 24 km/s, i.e., (144/n) km/s with $m = 1, 2, 3, \ldots$ This result, in particular the 72 km/s periodicity was confirmed by several authors [14, 15].

However, a global quantization with nearly the same velocity differences, 72 and 36 km/s [17], has also been found in samples on nearby galaxies, even when pairs are

Table 4 Velocity quantization of galaxy pairs

Double galaxy:	–	NGC-3958	NGC-3963	-	NGC-4294	NGC-4299	–	NGC-4085	NGC-4088	NGC-3504	NGC-3512	-	NGC-6542	NGC-6528
m_1 and m_2, $10^{10}(M_\odot)$	–	12.0	9.0	–	1.4	8.8	–	1.3	6.0	1.0	5.0	–	2.7	1.7
m	1		2	3		4	5		6		7	8.9		10
V_m (km s^{-1}) Theor.	–		72.3	–		36.1	–		24.1		20.6	–		14.4
Exp.	–		72	–		34	–		25		20	–		15

excluded. We shall now see how this effect can also be understood in our framework, but in terms of a cosmological effect, then recall briefly how the "global" quantization and the pair quantization must be related.

9 Atomic-Planetary Nebulae Analogies

Physically meaningful self-similarity (in any of its forms - prefractality, fractality or multifractality) does not require analogues on different scales to be exact replicas, but only to be similar in shape and motion, and related by consistent scaling relations [18]. Moreover, the atoms galaxies analogies find further support in the recent findings: that galaxies appear to have quantized redshifts, and that the statistics of galaxy clustering bear a remarkable ressemblance to results derived for subatomic particles. Galaxies and atoms share the same basic spherical, oblate and prolate shapes (spheroid, bipolar, propeller, caps, butterfly, toroid, sphere-within-a sphere, etc.) (Fig. 2).

For each atomic-planetary nebulae analogy, the left image is an atomic shape representing one of eight different excitations states (different fractal numbers n, l, m which determine the probability density distributions or really distributions of the physical mass of electron for a given energy state).

The images to the right are photographs and schematic models of planetary nebula star systems. Generally, the planetary nebulae are created when their central hot stars violently eject their outer shells. In this process, some loss of the presymmetry can be expected for these expanding astrophysical objects [19–22].

We present in Fig. 2 various analogies between atomic states - planetary nebulae according to (http://hubblesite.org/image/771/news/25-stellarjets; https://en.wikipedia.org/wiki/Abell39; https://en.wikipedia.org/wiki/CarinaNebula; http://demonstrations.wolfram.com/VisualizingAtomicOrbitals/).

10 Phase and Group Velocities, Fractal Type Uncertainty Relations and Their Implications

The hypothesis concerning the non-differentiability of the motion curves at any scale resolution has as an immediate consequence the fact that to any fractal type physical system (in a free motion on a fractal space - fractal type geodetic motion) one can associate at any scale resolution, a fractal type wave characterized through its state function Ψ,

$$\Psi(\mathbf{r}, t) = A \exp[-i(\omega t - \mathbf{kR})] \equiv A \exp[-\frac{i}{2M_0\lambda(dt)^{(2/D_F)-1}}(Et - \mathbf{PR})] \quad (58)$$

Fig. 2 Analogies between atomic states-planetary nebulae. Planetary nebula: Eta Carina, A-39, IRAS 04302+2247

with

$$E = MV_0^2 = 2M_0\lambda(dt)^{(2/D_F)-1}\nu$$
$$\mathbf{P} = 2M_0\lambda(dt)^{(2/D_F)-1}\mathbf{k} = M\mathbf{V} \tag{59}$$

where A is the wave amplitude, ω is the wave pulsation, ν is the wave frequency, $\mathbf{k}$ is the wave vector, E is the energy, $\mathbf{P}$ is momentum vector, $\mathbf{R}$ is the position vector, $\mathbf{V}$ is the velocity vector, t is the non-fractal time, M is the motion mass, $M = M_0(1 - V^2/V_0^2)^{-1/2}$, M_0 is the rest mass and V_0 is the limit velocity.

The propagation velocity of the wave can be considered to be equal to the propagation velocity of the wave surface (the phase velocity of the wave), V_f, i.e., of the surface for whom the phase in the case of the one-dimensional free motion

$$S = Et - PX = \text{const.} \tag{60}$$

i.e.,

$$V_f = \frac{dX}{dt} = \frac{E}{P} = \frac{MV_0^2}{MV} = \frac{V_0^2}{V} \tag{61}$$

Since $V < V_0$, the phase velocity is higher than the limit velocity V_0. The fact that the phase velocity V_f is higher than the limit velocity V_0 does not contradict the scale relativity principles [5, 6] because V_f is not a propagation velocity of the fractal type interactions.

Let us associate now, to any scale resolution and to any fractal type physical system (in a free motion on a fractal space), a waves package obtained by overlapping of an ensemble of plane harmonic waves, for which the momentum, P, of the fractal type physical system is contained in the interval

$$P_0 - \frac{\Delta P}{2} \leq P \leq P_0 + \frac{\Delta P}{2} \tag{62}$$

where P_0 is the momentum of the waves packet and ΔP is the momentum variation of the same package equipped by its "extension". Analogously with the quantum case one obtains:

$$\Psi(X, t) = A\exp[-\frac{i}{2M_0\lambda(dt)^{(2/D_F)-1}}(E_0T - P_0X)] \times \frac{\sin\xi}{\xi} \tag{63}$$

where

$$\xi = \frac{\Delta P}{4m_0\lambda(dt)^{(2/D_F)-1}}(X - \frac{\partial E}{\partial P}t) \tag{64}$$

Taking into account that the relation from the scale relativity between the wave package and the momentum for a free fractal type physical system has the form:

$$E = V_0\sqrt{M_0^2 V_0^2 + P^2}, \tag{65}$$

the group velocity of the wave package becomes:

$$V_G = \frac{\partial E}{\partial P} = \frac{V_0^2 P}{E} = \frac{V_0^2 M V}{M V_0^2} = V \tag{66}$$

In consequence, the group velocity of the wave package (66) is equal to the velocity of the fractal type physical system.

At a certain time t, the wave package (63) "extends" on the distance ΔX, obtained by the fractal relation:

$$\Delta\xi = \frac{\Delta P \Delta X}{2M_0\lambda(dt)^{(2/D_F)-1}} \geq \pi \tag{67}$$

or

$$\Delta P \Delta X \geq 2\pi M_0\lambda(dt)^{(2/D_F)-1} \tag{68}$$

The last expression corresponds to an uncertainty relation of fractal type. For the particular case of motions on Peano type fractal curves, $D_F = 2$, at Compton scale $\lambda = \hbar/M_0 c$, one obtains Heisenberg uncertainty relation:

$$\Delta P \Delta X \geq h. \tag{69}$$

Using a procedure similar to the one from [10], the fractal type uncertainty relation in energy-time becomes:

$$\Delta E \Delta t \geq 2\pi M_0\lambda(dt)^{(2/D_F)-1} \tag{70}$$

For the particular case of motions on Peano type fractal curves, $D_F = 2$, at Compton scale $\lambda = h/M_0 c$ one obtains the standard uncertainty relation:

$$\Delta E \Delta t \geq h \tag{71}$$

However, which are the implications if one admits an uncertainty relation at cosmological scale of the type

$$\Delta E \Delta t \geq h_G \tag{72}$$

where h_g is the gravitational type Planck constant, whose value is $h_G = (7-8)\cdot 10^{67}\,J\ s$ (for details see [20, 21]).

A numerical benchmark for the ground-state energy of a "typical" galaxies pair can be calculated by taking $m_1 \approx m_2 \approx 10^{41}$ kg. ($\approx$ the mass of the Milky Way). Then, through $E_{G_n} = -(1/2n^2)\mu c^2\alpha_g^2$, with μ the reduced mass, and n the main fractal number, for $n = 1$ one finds $E_1 = -10^{49}$ J. Using Eq. (72), the lifetime for excited states can be estimated. If ΔE is approximated as the energy difference between the ground state and the first excited state ($\approx |E_1|$), then, from Eq. (72) with $h_g \approx (7-8) \times 10^{67}$ J s [22], $\Delta t \approx 2 \times 10^{11}$ *year*. The excited state has a lifetime $\approx 10\times$(the age of Universe), as estimated using the standard cosmological model [22]. Therefore, if a double galaxy is formed in an excited state (or if it somehow reaches one after birth), it remains in that state for its entire existence.

The fractal excitation "energy" ($= h_g \nu$) is monochromatic gravitational radiation initially. It must be somehow converted to a spectrum of electromagnetic radiation, which is then emitted during the lifetime of the double galaxy. The conversion mechanism may be possibly related to gravitational collapse in the galactic nuclei of the pair. Since h_g is so large, it is plausible to expect radio wave emission to dominate, just as visible light dominates the emission spectrum of excited atoms (because h is so small). This consideration leads to two interesting possibilities which have not been considered:

(i) Double galaxies may be strong radio sources, compared to singles. Preliminary evidence supports this idea [20]. It was shown for a sample of isolated galaxies and isolated doubles, virtually all of them being spirals that compact radio sources occurred four times more frequently in doubles than in singles. There was also a higher occurence of strong radio emission from doubles compared to singles, with (on the average) a greater power output in double associated with the more frequent occurence of active galactic nuclei in their components. Although these results are encouraging, they are not definitive because of the small sample size and of other possible selection biases.

(ii) Some radio-active quasars may be cleverly disguised double galaxies in highly excited quantum states and at an early stage of evolution. Preliminary evidence suggests there may be localized radio sources inside some quasars, but better telescope resolution may be required to test the idea. However, there is some cause for cautious optimism. A rough estimate of the quasar's average power output in this model is of an acceptable order of magnitude, providing the mass of a "typical" component is taken as $m_1 \approx m_2 \approx 10^{43}$ kg. and the average time of existence is $\tau \approx 10^9$ year: i.e., $P_{\text{ave}} \approx (\varepsilon_n - \varepsilon_1)/\tau \approx 10^{39}$ J/s $\approx 10^6 L_\odot$, where $L_\odot$ is the Sun's luminosity.

11 Concluding Remarks

The main conclusions of this chapter are the following:

(i) It is built a dynamics of the multifractal type physical systems, considering that their entities simultaneously move on continuous but non-differentiable curves having different fractal dimensions. For this purpose, we firstly defined a fractal operator of motion with the role of scale covariant derivative and then, based on a scale covariance principles, we obtained the fractal space geodesics equations;

(ii) The separation of motions on scale resolutions (the differential one and the fractal one) in the geodesics equations induces a specific fractal force, which represents a measure of the non-differentiability of the motion curves;

(iii) The cancellation of the specific fractal force along with the satisfying of the states density conservation law at fractal scale resolutions imply the existence of a non-linear system of equations. In order to have a plane symmetry and adequate constraints (initial and to the limit), this system makes explicit a fractal velocities field in the form of the kink type, soliton type etc. solutions;

(iv) For irrotational motions of the physical system entities, we made explicit its geodesics in the form of a Schrödinger type fractal equation;

(v) Using the Schrödinger type fractal equation, we explicited similarities in the dynamics of certain physical systems at small scale resolutions,by means of atoms, respectively, at high scale resolutions by means of gravitational structures;

(vi) We proved that, at our solar system level, the dynamics (both of the planets around the Sun and of the natural satellites around the planets) show similarities in the form of certain quantifiable movements;

(vii) We demonstrated that the quantifiable movements at extra-galactic scale resolutions manifest in the form of red-sfiht quantization in binary galaxies. Moreover, atomic-planetary nebulae analogies were presented;

(viii) We obtained fractal uncertainty relations. In such framework, the admittance of an uncertainty relation at cosmological scale has a remarkable consequence in the sense that, for $h_g = (7 - 8) \times 10^{67}$ J s it results that a galaxy pair formed in an excited state of "hydrogen atom" type, remains in that state for its entire existence. The "quantum excitation energy" is monochromatic gravitational radiation initially. It must be somehow converted to a spectrum of electromagnetic radiation, which is then emitted during the lifetime of the double galaxy. The conversion mechanism may be possible related to gravitational collapse in the galactic nuclei of the pair. Since h_g is so large, it is plausible to expect radio wave emission to dominate, just as visible light dominates the emission spectrum of excited atoms (because h is so small).

References

1. B. Mandelbrot, *The Fractal Geometry of Nature* (W.H. Freeman Publishers, New York, 1982)
2. M.F. Barnsley, *Fractals Everywhere* (Morgan Kaufmann Publisher, San Francisco, 1993)
3. R. Badii, A. Politi, *Complexity: Hierarchical Structures and Scaling in Physics* (Cambridge University Press, Cambridge, 1997)
4. M. Mitchell, *Complexity: A Guided Tour* (Oxford University Press, Oxford, 2009)
5. L. Nottale, *Scale Relativity and Fractal Space-Time: A New Approach to Unifying Relativity and Quantum Mechanics* (Imperial College Press, London, 2011)
6. I. Mercheş, M. Agop, *Differentiability and Fractality in Dynamics of Physical Systems* (World Scientific Publisher, Singapore, 2016)
7. J. Cresson, Non-differentiable transformations on $\mathbb{R}^n$. Int. J. Geom. Meth. Mod. Phys. **03**, 13–45 (2006)
8. S. Florea, I. Dumitrache, *Elemente şi Circuite Fluidice (in Romanian)* (Romanian Academy Publishing House, Bucureşti, 1979)
9. P.C. Cristescu, *Dinamici neliniare şi haos. Fundamente teoretice şi aplicaţii (in Romanian)* (Romanian Academy Publishing House, Bucureşti, 2008)

10. S. Ţiţeica, *Quantum Mechanics* (Academic Press, Bucharest, 1984)
11. A.G. Agnese, R. Festa, Clues to discretization on the cosmic scale. Phys. Lett. A **227**, 165–171 (1997)
12. M. Mayor, D. Queloz, A Jupiter-Mass companion to a solar-type star. Nature **378**, 355–359 (1995)
13. Landolt-Börnstein, *Internal Structure and Dynamics of Galaxies*, Handbook VI/2C.9.2 (Springer, Berlin, 1982)
14. S.E. Schneider, E.E. Salpeter, Velocity differences in binary galaxies. I. Suggestions for a non-monotonic, two-component distribution. Astrophys. J. **385**, 32–48 (1992)
15. W.J. Cocke, Statistical methods for investigating periodicities in double-galaxy redshifts. Astrophys. J. **393**, 59–67 (1992)
16. W.G. Tifft, W.J. Cocke, Global redshift quantization. Astrophys. J. **287**, 492–507 (1984)
17. B.N. Guthrie, W.M. Napier, Mon. Not. R. Astron. Soc. **253**, 533 (1991)
18. J. Argyris, C. Marin, C. Ciubotariu, *Physics of Gravitation and the Universe* (Spiru Haret Publishing House, Iaşi, 2002 and Tehnica-Info Publishing House, Chişinău, Vol. II, 2002)
19. H.E. White, Pictorial representations of the electron cloud for hydrogen-like atoms. Phys. Rev. **37**, 1416 (1931)
20. M. DerSarkissian, Does wave-particle duality apply to galaxies? Lett Nuovo Cimento **40**(13), 390–394 (1984)
21. M. Agop, H. Matsuzawa, I. Oprea, R. Vlad, C. Sandu, CGh Buzea, Some implications of the gravitomagnetic field in fractal spacetime theory. Aust. J. Phys. **53**, 217 (1999)
22. S. Weinberg, *Gravitation and Cosmology: Principles and Applications of the General Theory of Relativity* (Wiley, New York, 1972)

Plasma Perturbations and Cosmic Microwave Background Anisotropy in the Linearly Expanding Milne-Like Universe

S. L. Cherkas and V. L. Kalashnikov

1 Introduction

Present universe is transparent for photons, but it was not the same before the hydrogen recombination at the red-shifts of $z \approx 1100$,[1] when it was filled with the photon-baryon plasma. Protons and electrons were coupled to the radiation through the Compton scattering by electrons which in turn are coupled to the baryons by Coulomb interaction [1, 3]. Such primordial plasma perturbations were widely considered in cosmology, and their fingerprints depend on a law of the universe expansion that is the crucial point for our further analysis.

Recently, the Milne-like cosmologies considering the linearly expanding (in cosmic time) universe models [4, 5] again attract an attention [6–15]. Instead of the

[1] z is the red-shift parameter used as a measure of cosmological time and distance: $z + 1 = a_0/a(\eta)$, where a_0 is the present scale factor value, and $a(\eta)$ is the scale factor at some earlier photon emission time η [1, 2].

S. L. Cherkas
Institute for Nuclear Problems, Bobruiskaya str. 11,
Minsk 220050, Belarus
e-mail: cherkas@inp.bsu.by

V. L. Kalashnikov (✉)
Institute of Photonics, Vienna University of Technology,
Vienna 1040, Austria
e-mail: vladimir.kalashnikov@tuwien.ac.at

C. H. Skiadas (ed.), *Fractional Dynamics, Anomalous Transport and Plasma Science*, https://doi.org/10.1007/978-3-030-04483-1_9

original open and empty Milne universe model [4, 5],[2] the flat universes filled with some exotic matter are considered. It seems reasonable to associate such "a primordial matter fluid" with the vacuum [16].

We will consider the perturbations of plasma consisting of photons, baryons, and electrons in a linearly expanding (Milne-like) universe with taking into account the metric tensor and vacuum perturbations. Here, we will use the oversimplified model of plasma as a *pure radiation*, i.e., a substance with the equation of state $w = 1/3$,[3] to obtain an analytical solution. This approximation is admissible because initially, the temperature is sufficiently large to consider all the particles as a relativistic fluid. Then, the particles decay eventually to the photons, electrons, and baryons. According to observations number of photons is of 10^9 times larger than that of nucleons and electrons. Thus, the nucleons contribute at only the late stage of the universe evolution. We will base our analysis of the metric tensor perturbations, which contribute to the primordial plasma formation, on the five-vectors theory of gravity [18]. The quantization of this model could resolve the problem of huge vacuum energy [19] and allow omitting its main part.[4]

2 Perturbations of Plasma and Vacuum

We expose the perturbation theory for primordial photon-baryon plasma, vacuum and metric tensor. Vacuum issue is the well-known challenge for quantum or, at least, semiclassical theory [16, 19–22]. Here, we will consider a vacuum purely classically, that is as a substance producing the linear expansion of the universe in the framework of the developed theory [18] which admits adding or extracting some constant to the energy density.

2.1 Underlying Gravity Theory

The conventional theory of the CMB spectrum is the General Relativity theory (GR) (e.g., see [2]). In the case of the Milne-like cosmology, the issue is more complicated, because an origin of linear universe expansion is not clear. As was shown, such linear expansion could arise from the residual vacuum fluctuations of quantized fields including the scalar and gravitational ones after omitting the main part of huge vacuum energy [16]. As was mentioned above, the mystery of cosmological vacuum

---

[2]The universe proposed initially by Milne describes an open and empty (i.e., Minkowski) spacetime which expands linearly with time [1, 4, 5]. It is negatively-curved spatially (i.e., hyperbolical) in 3-dimensions but is "flat" in 4 (i.e., spacetime)-dimensions.

[3]We use a classical definition for the equation of state parameter w corresponding to a perfect fluid, that is the ratio of pressure to density [17].

[4]Below, the system of units $\hbar = c = 1$ will be used, and we define the present scale factor as $a_0 = 1$.

is among the critical issues of modern physics [19, 21, 22]. Below we will use the theory which validates omitting the vacuum extra-energy and, besides, provides obtaining the analytical solutions.

Let's start from the Einstein-Hilbert action for GR in the form of [23]:

$$S = -\frac{M_p^2}{12}\int \mathcal{G}\sqrt{-g}\,d^4x, \tag{1}$$

where $\mathcal{G} = g^{\alpha\beta}\left(\Gamma^{\rho}_{\alpha\nu}\Gamma^{\nu}_{\beta\rho} - \Gamma^{\nu}_{\alpha\beta}\Gamma^{\rho}_{\nu\rho}\right)$, and M_p is the Planck mass, which is chosen as $M_p = \sqrt{\frac{3}{4\pi G}}$.

The next step is a violation of the general coordinate covariance principle in (1) according to the Milne's perception of the principally different concept of time in GR and quantum mechanics [24–26], so that we will consider the restricted class of metrics $g_{\mu\nu}$ in the form of

$$ds^2 \equiv g_{\mu\nu}dx^\mu dx^\nu = a^2\left(1-\partial_m P^m\right)^2 d\eta^2 - \gamma_{ij}(dx^i + N^i d\eta)(dx^j + N^j d\eta), \tag{2}$$

where γ_{ij} is the induced three metric, $a = \gamma^{1/6}$ is the scale factor defined locally, and $\gamma = \det\gamma_{ij}$. A spatial part of the interval (2) can be written as

$$dl^2 \equiv \gamma_{ij}dx^i dx^j = a^2(\eta, \boldsymbol{x})\tilde{\gamma}_{ij}dx^i dx^j, \tag{3}$$

where $\tilde{\gamma}_{ij} = \gamma_{ij}/a^2$ is a matrix with the unit determinant. The interval (2) is analogous to the the ADM one [27], but the expression $1 - \partial_m P^m$ is used instead of a lapse function, where ∂_m is a partial derivative and P^m is a three-vector. Varying the action over vectors $\boldsymbol{P}$, $\boldsymbol{N}$ and three metric γ_{ij} leads to the equations of the five-vectors theory (FVT) [18]:

$$\begin{gathered}
\frac{\partial g^{\mu\nu}}{\partial\gamma_{ij}}\left(\frac{(\partial\mathcal{G}\sqrt{-g})}{\partial g^{\mu\nu}} - \frac{\partial}{\partial x^\lambda}\frac{\partial(\mathcal{G}\sqrt{-g})}{\partial(\partial_\lambda g^{\mu\nu})} - \frac{6}{M_p^2}T_{\mu\nu}\sqrt{-g}\right) = 0,\\
\frac{\partial g^{\mu\nu}}{\partial N^i}\left(\frac{\partial(\mathcal{G}\sqrt{-g})}{\partial g^{\mu\nu}} - \frac{\partial}{\partial x^\lambda}\frac{\partial(\mathcal{G}\sqrt{-g})}{\partial(\partial_\lambda g^{\mu\nu})} - \frac{6}{M_p^2}T_{\mu\nu}\sqrt{-g}\right) = 0,\\
\frac{\partial g^{\mu\nu}}{\partial(\partial_j P^i)}\frac{\partial}{\partial x^j}\left(\frac{\partial(\mathcal{G}\sqrt{-g})}{\partial g^{\mu\nu}} - \frac{\partial}{\partial x^\lambda}\frac{\partial(\mathcal{G}\sqrt{-g})}{\partial(\partial_\lambda g^{\mu\nu})} - \frac{6}{M_p^2}T_{\mu\nu}\sqrt{-g}\right) = 0.
\end{gathered} \tag{4}$$

Eq. (4) are *weaker* than the GR ones. At the same time, the restrictions $\nabla(\nabla\cdot\boldsymbol{P}) = 0$ and $\nabla(\nabla\cdot\boldsymbol{N}) = 0$ on the Lagrange multipliers arise [18]. In the particular case of $\nabla\cdot\boldsymbol{N} = 0$, *the Hamiltonian constraint is satisfied up to some constant.*

The next step is to develop a theory for the scalar perturbations in the gauge of $\boldsymbol{P} = 0$, $\boldsymbol{N} = 0$:

$$ds^2 = a(\eta)^2(1+2A)\left(d\eta^2 - \left(\left(1+\frac{1}{3}\sum_{m=1}^{3}\partial_m^2 F\right)\delta_{ij} - \partial_i\partial_j F\right)dx^i dx^j\right). \quad (5)$$

An interval (5) is a particular form of the interval (2) up to the higher order terms in $F(\eta, \boldsymbol{x})$ by virtue of

$$\ln\left[\det\left(\left(1+\frac{1}{3}\sum_m \partial_m^2 F\right)\delta_{ij} - \partial_i\partial_j F\right)\right] \approx \operatorname{tr}\left(\left(\frac{1}{3}\sum_m \partial_m^2 F\right)\delta_{ij} - \partial_i\partial_j F\right) = 0.$$

Writing Eq. (4) up to the first order relatively $A(\boldsymbol{x}, \eta)$ and $F(\boldsymbol{x}, \eta)$ leads to the required perturbation theory.

2.2 Energy-Momentum Tensor

As was above mentioned, we violate eventually the general coordinates' transformation invariance by the restriction of the metrics' class by representing them in the form of (2). To built the energy-momentum tensor in the field theory, one should write the corresponding special relativistic expression and then change the partial derivatives to covariant ones. Using a hydrodynamic approximation is more convenient. In this framework the energy-momentum tensor is

$$T_{\mu\nu} = (p+\rho)u_\mu u_\nu - p\, g_{\mu\nu}. \quad (6)$$

The equations of motion for some fluid in the GR can be obtained from both the equations of motion of the fluid point-like components and the conservation of the energy-momentum tensor $D_\mu T^{\mu\nu} = 0$, where D_μ is a covariant derivative. In FVT, the energy-momentum tensor conserves only in the Minkowski space-time. However, one can deduce the equation of motion for fluid from the conservation of energy-momentum tensor by virtue of the Eq. (6) self-consistency in the particular gauge (5). Below, we will consider the scalar perturbations of a fluid c (the index c denotes a kind of fluid) in the form of $\rho_c(\eta, \boldsymbol{x}) = \rho_c(\eta) + \delta\rho_c(\eta, \boldsymbol{x})$, $p_c(\eta, \boldsymbol{x}) = p_c(\eta) + \delta p_c(\eta, \boldsymbol{x})$ and represent the 4-velocity in the form of

$$u_c^\mu = \frac{1}{a(\eta)}\{(1-A), \boldsymbol{\nabla} v_c(\eta, \boldsymbol{x})\}, \quad (7)$$

where $v_c(\eta, \boldsymbol{x})$ is a scalar function.

2.3 Zero-Order Equations

The zero-order evolution equation for logarithm of the scale factor $\alpha(\eta) = \ln a(\eta)$ takes the form of

$$\alpha'' + \alpha'^2 = M_p^{-2} e^{2\alpha}(\rho - 3p), \tag{8}$$

where $\rho = \sum_c \rho_c$ and $p = \sum_c p_c$ are the uniform energy density and pressure, respectively. Summation is performed over all the kinds of matter, but here we will consider only vacuum $c = v$ and radiation $c = r$. For every component of a substance, the equation of motion is:

$$\rho_c' + 3\alpha'(\rho_c + p_c) = 0. \tag{9}$$

Pressure of a fluid is connected with the energy density as $p_c = w_c \rho_c$ (see the footnote 3 above and Ref. [17]). It is worth mentioning that the Friedmann equation is satisfied only up to some constant in the framework of the model considered:

$$M_p^{-2} e^{4\alpha} \rho(\eta) - \frac{1}{2} e^{2\alpha} \alpha'^2 = const, \tag{10}$$

that is the integral of motion of Eqs. (8), (9).

As was shown [28], the residual vacuum fluctuations can explain a nearly-linear universe expansion. Here, for simplicity, we will use an empirical consideration. Let us analyze a linear universe expansion that means $a(\eta) = B \exp(\mathcal{H}\eta)$ in conformal time, and find the corresponding empirical equation for the vacuum state. The very simple equation of state arises if we set a constant in the Friedmann equation (10) so that

$$M_p^{-2} e^{4\alpha} \rho_v - \frac{1}{2} e^{2\alpha} \alpha'^2 = 0. \tag{11}$$

It is possible because $\rho_r e^{4\alpha}$ is also constant. Under such choice of a constant, the equation of the vacuum state will be $w_v = -1/3$. This equation of state is widely discussed earlier [9, 10, 14]. One may obtain from Eq. (9) $\rho_v e^{2\alpha} = const$ for the vacuum, that results in (see Eq. 11):

$$a(\eta) = \exp(\alpha(\eta)) = B \exp(\mathcal{H}\eta), \tag{12}$$

where B is some constant. In the cosmic time $dt = a(\eta) d\eta$

$$a(t) = \mathcal{H} t, \tag{13}$$

i.e., it is a linear expansion of the universe.

2.4 Perturbations

Introducing the quantity $V_c = (p_c + \rho_c)v_c$ for every fluid c and expanding all perturbations into the Fourier series $\delta\rho_c(\boldsymbol{x}) = \sum_k \delta\rho_{ck} e^{i\boldsymbol{kx}}$... etc. result in the equations for perturbations:

$$-6A'_k + 6A_k\alpha' + k^2 F'_k + \frac{18}{M_p^2} e^{2\alpha} \sum_c V_{ck} = 0, \quad (14)$$

$$-18\alpha' A'_k - 18A_k\alpha'^2 - 6k^2 A_k + k^4 F_k + \frac{18}{M_p^2} e^{2\alpha} \sum_c \delta\rho_{ck} + 4A_k\,\rho_c = 0, \quad (15)$$

$$-12A_k - 3\left(F''_k + 2\alpha' F'_k\right) + k^2 F_k = 0, \quad (16)$$

$$-9\left(A''_k + 2\alpha' A'_k\right) - 18A_k\alpha'' - 18A_k\alpha'^2 - 9k^2 A_k + k^4 F_k$$
$$-\frac{9}{M_p^2} e^{2\alpha} \sum_c 4A_k(3p_c - \rho_c) + 3\delta p_{ck} - \delta\rho_{ck} = 0, \quad (17)$$

$$-3\alpha'(\delta p_{ck} + \delta\rho_{ck}) - 3A'_k(\rho_c + p_c) - \delta\rho'_{ck} + k^2 V_{ck} = 0, \quad (18)$$

$$(\rho_c + p_c)A_k + 4V_{ck}\alpha' + \delta p_{ck} + V'_{ck} = 0. \quad (19)$$

The last two equations, obtained from the energy-momentum conservation, are assumed to be valid for every c-substance under consideration. The choice of the constant in Eq. (10) is arbitrary. The constraint Eqs. (14) and (15) are consistent with other equations under this arbitrary choice. It is not true in a perturbation theory within the framework of GR, where a perturbation of the constraint equations is consistent with other equations only if a sum of the mean densities of all fluids equals the critical density (for the flat universe). Here we consider the flat universe in a mean, but the sum of the mean densities is determined up to some constant, and nevertheless, all the equations for perturbations are self-consistent. With that chosen constant in Eq. (10), the radiation does not affect the universe expansion and the equation of state $w_v = -1/3$ for the vacuum results in linear expansion of the universe. Thus, the equations of state are $w_v = -1/3$ for the vacuum and $w_r = 1/3$ for the radiation.[5]

Such choice of the constant in (10) is an invention expired by the existence of the analytical solution in this case. The above system of the equations can be reduced to a single linear equation with the constant coefficients under the assumption of $a(\eta) = B\exp(\mathscr{H}\eta)$ and $\rho_r = \frac{\rho_{r0}}{a^4(\eta)}$, where ρ_{r0} is a density of radiation at the present time:

$$9\delta\rho_{rk}^{(4)} + 6\left(30\mathscr{H}\delta\rho_{rk}^{(3)} + \left(222\mathscr{H}^2 + k^2\right)\delta\rho''_{rk} + 10\mathscr{H}\left(72\mathscr{H}^2 + k^2\right)\delta\rho'_{rk}\right)$$
$$+\left(48\mathscr{H}^2 + k^2\right)\left(108\mathscr{H}^2 + k^2\right)\delta\rho_{rk} = 0. \quad (20)$$

---

[5] $\delta p_{ck} = w_c\delta\rho_{ck}$ is assumed, as well.

That allows obtaining the solution for the perturbation of radiation density:

$$\delta\rho_{r\boldsymbol{k}} = \mathrm{e}^{-6\eta\mathscr{H}}\left(C_1 \mathrm{e}^{-i\frac{\eta k}{\sqrt{3}}} + C_2 \mathrm{e}^{i\frac{\eta k}{\sqrt{3}}}\right) + \mathrm{e}^{-4\eta\mathscr{H}}\left(C_3 \mathrm{e}^{-i\frac{\eta k}{\sqrt{3}}} + C_4 \mathrm{e}^{i\frac{\eta k}{\sqrt{3}}}\right). \quad (21)$$

For a "flux" of the radiation fluid $V_{r\boldsymbol{k}}$, we have

$$V_{r\boldsymbol{k}} = \frac{B^2\mathscr{H}{M_p}^2}{6k\rho_{r0}}\mathrm{e}^{-4\eta\mathscr{H}}\left(C_1\left(k - i\sqrt{3}\mathscr{H}\right)\mathrm{e}^{-i\frac{\eta k}{\sqrt{3}}} + C_2\left(k + i\sqrt{3}\mathscr{H}\right)\mathrm{e}^{\frac{i\eta k}{\sqrt{3}}}\right). \quad (22)$$

Other functions $A_{\boldsymbol{k}}$, $F_{\boldsymbol{k}}$, $\delta\rho_{v\boldsymbol{k}}$, $V_{v\boldsymbol{k}}$ found from the system (14)–(19) are presented in Appendix.

The constants C_1, C_2, C_3, C_4 have to be determined from the initial conditions. The constants Z_1, Z_2 (see Appendix) do not contribute to the radiation density perturbations. Thus, we will equal them to zero. Indeed, it is reasonable to assume that an empty universe (i.e., filled by the only vacuum) has no any rising physical perturbation, and only perturbations connected with the radiation over the vacuum have a physical meaning. For simplicity, we assume that the only perturbations of radiation density $\delta\rho_{r\boldsymbol{k}}(\eta_{in})$ are non-zero initially, where η_{in} is an initial moment in conformal time.

Then, the solutions of the perturbation theory equations take the form:

$$\delta\rho_{r\boldsymbol{k}}(\eta) = \mathrm{e}^{4\mathscr{H}(\eta_{in}-\eta)}\left(4\sqrt{3}\mathscr{H}\sin\left(\frac{k(\eta-\eta_{in})}{\sqrt{3}}\right)\right. \quad (23)$$

$$\left.+k\cos\left(\frac{k(\eta-\eta_{in})}{\sqrt{3}}\right)\right)\delta\rho_{r\boldsymbol{k}}(\eta_{in})/k, \quad (24)$$

$$V_{r\boldsymbol{k}}(\eta) = 0, \qquad A_{\boldsymbol{k}}(\eta) = -\frac{B^4\mathrm{e}^{4\eta\mathscr{H}}}{4\rho_{r0}}\delta\rho_{r\boldsymbol{k}}(\eta), \quad (25)$$

$$F_{\boldsymbol{k}}(\eta) = -\frac{3B^4\mathrm{e}^{4\mathscr{H}\eta_{in}}}{2k^2\rho_{r0}\left(3\mathscr{H}^2+k^2\right)}\left(\left(12\mathscr{H}^2+k^2\right)\cos\left(\frac{k(\eta-\eta_{in})}{\sqrt{3}}\right)\right.$$

$$\left.+3\sqrt{3}\mathscr{H}k\sin\left(\frac{k(\eta-\eta_{in})}{\sqrt{3}}\right)\right)\delta\rho_{r\boldsymbol{k}}(\eta_{in}), \quad (26)$$

$$V_{v\boldsymbol{k}}(\eta) = \frac{B^2\mathscr{H}^2{M_p}^2\mathrm{e}^{4\mathscr{H}\eta_{in}-2\eta\mathscr{H}}}{12k\rho_{r0}\left(3\mathscr{H}^2+k^2\right)}\left(\sqrt{3}\left(12\mathscr{H}^2+k^2\right)\sin\left(\frac{k(\eta-\eta_{in})}{\sqrt{3}}\right)\right.$$

$$\left.-9\mathscr{H}k\cos\left(\frac{k(\eta-\eta_{in})}{\sqrt{3}}\right)\right)\delta\rho_{r\boldsymbol{k}}(\eta_{in}), \quad \delta\rho_{v\boldsymbol{k}}(\eta) = 3\mathscr{H}V_{v\boldsymbol{k}}(\eta). \quad (27)$$

The quantities $V_{v\boldsymbol{k}}(\eta)$ and $\delta\rho_{v\boldsymbol{k}}(\eta)$ will not be needed for the CMB spectrum calculations and will not be considered further.

2.5 "Gauge Invariant" Variables

The issue is that the metric (5) has not a typical form of

$$ds^2 = a^2(\eta)\left((1+2\Phi(\eta,\boldsymbol{x}))d\eta^2 - (1-2\Psi(\eta,\boldsymbol{x}))\,\delta_{ij}dx^i dx^j\right), \tag{28}$$

which appears in the conventional perturbation theory [1, 3] of GR. The comparability of previous results with those of the GR conventional perturbation theory can be provided by the "gauge invariant" densities, velocities and potentials [1]:

$$\begin{aligned}
\tilde{\delta}_{rk}(\eta) &= \frac{\delta\rho_{rk}(\eta)}{\rho_r(\eta)} - 2\alpha'(\eta)F_k'(\eta), \quad \tilde{v}_{rk} = \frac{V_{rk}(\eta)}{\rho_r(\eta)+p_r(\eta)} - \frac{f_k'(\eta)}{2},\\
\Phi_k(\eta) &= A_k(\eta) + \frac{a'(\eta)F_k'(\eta) + a(\eta)F_k''(\eta)}{2a(\eta)},\\
\Psi_k(\eta) &= -\frac{a'(\eta)F_k'(\eta)}{2a(\eta)} - A_k(\eta) + \frac{1}{6}k^2F_k(\eta).
\end{aligned} \tag{29}$$

We could not work with the "invariant" potentials initially because the metric (28) has not the form (2) and does not admit obtaining the consistent system of the equations when the zero-order Friedmann equation is violated, i.e., satisfied up to some constant (10). For our simplified approach, when only initial value of $\delta\rho_{rk}$ is nonzero, the calculated "invariant quantities" are

$$\begin{aligned}
\tilde{\delta}_{rk}(\eta) &= \frac{1}{\left(3\mathscr{H}^2+k^2\right)}\left(\left(12\mathscr{H}^2+k^2\right)\cos\left(\frac{k(\eta-\eta_{in})}{\sqrt{3}}\right)\right.\\
&\quad\left. +3\sqrt{3}\mathscr{H}k\sin\left(\frac{k(\eta-\eta_{in})}{\sqrt{3}}\right)\right)\delta_{rk}(\eta_{in}),\\
\tilde{v}_{rk}(\eta) &= \frac{1}{4k\left(3\mathscr{H}^2+k^2\right)}\left(9\mathscr{H}k\cos\left(\frac{k(\eta-\eta_{in})}{\sqrt{3}}\right)\right.\\
&\quad\left. -\sqrt{3}\left(12\mathscr{H}^2+k^2\right)\sin\left(\frac{k(\eta-\eta_{in})}{\sqrt{3}}\right)\right)\delta_{rk}(\eta_{in}),\\
&\Phi_k(\eta) = 0, \quad \Psi_k(\eta) = 0,
\end{aligned} \tag{30}$$

where we take into account that $\frac{\rho_{r0}}{B^4\exp(4\mathscr{H}\eta_{in})} = \rho_r(\eta_{in})$ and $\delta_{rk}(\eta_{in}) = \frac{\delta\rho_{rk}(\eta_{in})}{\rho_r(\eta_{in})}$. The potentials Φ_k, Ψ_k are zero only because we use the simplified initial condition, where $\delta\rho_{rk}$ is nonzero initially.

2.6 *Silk Dumping*

Electrons scatter the photons before the time of the last scattering surface. Although we consider photon-electron-baryon plasma as some perfect medium with the equation of state $w = 1/3$, the photon diffusion due to the Thompson scattering exists [2]. To estimate this (so-called *Silk dumping*) contribution to the perturbations, we follow the methodology of Refs. [1, 3] suggesting the suppression of the expressions (23), (25)–(27) and (29) by the factor $\exp\left(-k^2/k_D^2\right)$, where k_D is written as [1]

$$k_D(\eta_r) \approx \left(\frac{2}{15}\int_0^{\eta_r}\frac{d\eta}{\sigma_T n_e a}\right)^{-1/2} = \left(\frac{2}{15\sigma_T n_{b0}}\int_0^{\eta_r} a^2 d\eta\right)^{-1/2}, \tag{31}$$

and $\sigma_T = 6.65\times 10^{-25}\ \mathrm{cm}^2$ is the Thompson cross section. The free electron density n_e before recombination equals to the baryon density and scales as $n_e = n_{b0}a^{-3}$, where n_{b0} is the baryon present density

$$n_{b0} = \Omega_b \frac{M_p^2 \mathscr{H}^2}{2m_p} \tag{32}$$

expressed through a dimensionless quantity Ω_b, a proton mass m_p and a critical density $M_p^2\mathscr{H}^2/2$. Formally, for the dependence given by (12), an integration in (31) has to begin from $\eta = -\infty$. However, as was shown in [28], the universe started from a power-law expansion changed by (12) afterward. It was also shown, that B is of the order of 10^{-30}. Under this condition, B does not play a role if the lower limits of η equal $-\infty$ or zero (the results are approximately the same in both cases).

Substituting the dependence (12) and the conformal time of the last scattering surface $\eta_r = \frac{1}{\mathscr{H}}\ln\frac{10^{-3}}{B}$, that corresponds to the scale factor $a_r \approx 10^{-3}$, into (31) results in

$$k_D(\eta_r) = \sqrt{15\sigma_T n_{b0}\mathscr{H}} \times 10^3 \approx 10^3\sqrt{\Omega_b}\,\mathscr{H}. \tag{33}$$

As one may see, plasma is closer to an ideal fluid for greater matter density. For instance, the conventional value of $\Omega_b = 0.03$ results in the damping scale of $k_D \sim 170$ in the units of $\mathscr{H}$.

3 CMB Spectrum

In the previous section, we have considered the perturbation theory which describes the evolution of the plasma (radiation) in the presence of the vacuum perturbations. This evolution extends up to the "last scattering surface", i.e., up to a moment when the universe becomes transparent for radiation. Conformal time of the last scattering surface η_r corresponds to the temperatures $T_r \sim 3000\ K$ and the redshift $z_r \approx 1100$. Describing the photons' propagation from the last scattering surface to an observer is insufficient to use hydrodynamic approximation so that the Boltzmann equation is needed, which can be written in the form of

$$\frac{\partial f}{\partial \eta} + \frac{dx^i}{d\eta}\frac{\partial f}{\partial x^i} + \frac{dp_i}{d\eta}\frac{\partial f}{\partial p_i} = St[f], \tag{34}$$

where the right hand side $St[f]$ represents the collision integral. If the distribution function f is assumed to be a scalar, it would depend on x^i and p_i because the photon number $dN = f(x^i, p_j, \eta)dx^1dx^2dx^3dp_1dp_2dp_3$ is scalar according to the Liouville theorem and the quantity $dx^1dx^2dx^3dp_1dp_2dp_3$ is scalar. The expressions describing the photon propagation are

$$\frac{dp^\alpha}{d\lambda} = -\Gamma^\alpha_{\beta\gamma} p^\beta p^\gamma = -\Gamma^\alpha_{\beta\gamma} g^{\beta\sigma} g^{\gamma\delta} p_\sigma p_\delta,$$
$$\frac{dx^\alpha}{d\lambda} = p^\alpha = g^{\alpha\beta} p_\beta, \tag{35}$$

where λ is an affine parameter along the photon trajectory. Using the last equation for the zero component $\frac{dx^0}{d\lambda} = \frac{d\eta}{d\lambda} = p^0$ of derivatives with respect to λ allows rewriting it in the terms of derivatives with respect to η.

Then, the Boltzmann equation can be reduced to the equation for a temperature perturbation by substitution

$$f(x^i, p_j, \eta) = \frac{1}{\exp\left(\frac{p_0(\eta)}{T_0(\eta)\sqrt{g_{00}}(1+\Theta(\boldsymbol{n},\boldsymbol{x},\eta))}\right) - 1}, \tag{36}$$

where $\Theta(\boldsymbol{n}, \boldsymbol{x}, \eta)$ is a temperature contrast and a unit vector $n^i = p_i/(\sum_{n=1}^3 p_n^2)$. Finally, for the coefficients of the Fourier transform $\Theta(\boldsymbol{n}, \boldsymbol{x}, \eta) = \sum_{\boldsymbol{k}} \Theta_{\boldsymbol{k}}(\eta, \boldsymbol{n})\mathrm{e}^{i\boldsymbol{k}\boldsymbol{x}}$ calculations with the metric (5) give

$$\frac{\partial \Theta_{\boldsymbol{k}}}{\partial \eta} - ik\mu\Theta_{\boldsymbol{k}} - ik\mu A_{\boldsymbol{k}} + A'_{\boldsymbol{k}} + \frac{k^2}{6}\left(3\mu^2 - 1\right) F'_{\boldsymbol{k}} = \tau'(\Theta_{\boldsymbol{k}} - \Theta_{0\boldsymbol{k}} - v_{b\boldsymbol{k}}\,\mu), \tag{37}$$

where $\mu = \boldsymbol{n}\cdot\boldsymbol{k}/k$ is the cosine of the angle between $\boldsymbol{n}$ and $\boldsymbol{k}$, $\Theta_{0\boldsymbol{k}}(\eta)$ is the component $l = 0$ of $\Theta_{\boldsymbol{k}}(\boldsymbol{n}, \eta)$ in the expansion of the Legendre polynomials

$$\Theta_{l\boldsymbol{k}} = i^l \int_{-1}^{1} P_l(\mu)\Theta_{\boldsymbol{k}}(\mu)\frac{d\mu}{2}, \tag{38}$$

and $v_{b\boldsymbol{k}}$ is the Fourier transform of the function determining baryon velocity. The function $\tau(\eta)$ describes the photon Compton scattering by electrons: $\tau' = -\sigma_T n_e a$, where σ_T is a cross section of the Thomson scattering and n_e is a free electron density. Before the last scattering surface, the photons are tightly coupled with electrons and protons by the Thomson scattering, and the electrons, in turn, are tightly coupled with baryons by the Coulomb interaction. As a consequence, any bulk motion of the photons must be shared by the baryons. Although we do not consider baryons explicitly, one may assume roughly that baryons and photons are in equilibrium and thus [3]

$$v_{b\boldsymbol{k}} = -3i\Theta_{1\boldsymbol{k}}(\eta). \tag{39}$$

Further, the monopole $\Theta_{0\boldsymbol{k}}$ and dipole $\Theta_{1\boldsymbol{k}}$ components of the temperature perturbations can be connected with the perturbations of density and velocity. From one hand side, the 00-component of the energy-momentum tensor in line with (6) is

$$T^0_{0\boldsymbol{k}} = \delta\rho_{\boldsymbol{k}}(\eta). \tag{40}$$

On the other hand, it can be expressed via a temperature perturbation [1]:

$$T^0_{0\boldsymbol{k}} = 4\rho_r \int \Theta_{\boldsymbol{k}}(\boldsymbol{n}, \eta)\frac{d^2\boldsymbol{n}}{4\pi}. \tag{41}$$

Comparison of (40) and (41) gives $\Theta_{0\boldsymbol{k}}(\eta) = \frac{1}{4\rho_r}\delta\rho_{r\boldsymbol{k}}(\eta) = \frac{1}{4}\delta_{r\boldsymbol{k}}$. Analogously, in the first order of the perturbation theory, the components T_{0i} take the form of

$$T_{0j} = -a^2(\eta)(\rho_r(\eta) + p_r(\eta))\partial_j v_r(\eta, \boldsymbol{x}), \tag{42}$$

or

$$T^j_{0\boldsymbol{k}} = \frac{4}{3}\rho_r(\eta) i k_j v_{r\boldsymbol{k}}(\eta). \tag{43}$$

At the same time [1]

$$T^j_{0\boldsymbol{k}} = -4\rho_r \int n^j \Theta_{\boldsymbol{k}}(\boldsymbol{n}, \eta)\frac{d^2 n}{4\pi}. \tag{44}$$

As consequence of (38), (39), (43) and (44), one has $v_{b\boldsymbol{k}} = -3i\Theta_{1\boldsymbol{k}} = -ikv_{r\boldsymbol{k}}$, and Eq. (37) can be rewritten in the form of

$$\Theta'_{\boldsymbol{k}} - (ik\mu + \tau')\Theta_{\boldsymbol{k}} = \mathrm{e}^{ik\mu\eta+\tau}\frac{d}{d\eta}\left(\Theta_{\boldsymbol{k}}\mathrm{e}^{-ik\mu\eta-\tau}\right) - S_{\boldsymbol{k}}, \tag{45}$$

where $S_{\boldsymbol{k}} = -\tau'\frac{\delta_{r\boldsymbol{k}}}{4} + \tau' ik\mu\, v_{r\boldsymbol{k}} + ik\mu A_{\boldsymbol{k}} - A'_{\boldsymbol{k}} - \frac{k^2}{6}\left(3\mu^2 - 1\right)F'_{\boldsymbol{k}}$.

Solution of Eq. (45) takes the form of

$$\Theta_k(\eta_0) = \Theta_k(\eta_{in})\mathrm{e}^{-i\mu k(\eta_{in}-\eta_0)-\tau(\eta_{in})+\tau(\eta_0)} + \int_{\eta_{in}}^{\eta_0} S_k\, \mathrm{e}^{-i\mu k(\eta-\eta_0)-\tau(\eta)+\tau(\eta_0)}d\eta \approx$$
$$\int_{\eta_{in}}^{\eta_0} \mathrm{e}^{-\tau(\eta)}\Bigg(-\tau'\frac{\delta_{rk}}{4} - \tau' v_{rk}\frac{d}{d\eta} - A'_k - A_k\frac{d}{d\eta}$$
$$-\frac{F'_k}{6}\left(-3\frac{d^2}{d\eta^2} - k^2\right)\Bigg)\mathrm{e}^{-ik\mu(\eta-\eta_0)}d\eta, \quad (46)$$

where η_0 is the present day conformal time, η_{in} is some initial moment of time before the last scattering surface, when the universe was not transparent for light. The terms containing $\mathrm{e}^{-\tau(\eta_{in})}$ are omitted because the function $\mathrm{e}^{-\tau(\eta)}$ vanishes quickly if $\eta < \eta_r$ [3].

Using the integral (38) and the integral

$$\int_{-1}^{1}\frac{d\mu}{2}P_l(\mu)\mathrm{e}^{-ik\mu(\eta-\eta_0)} = \frac{1}{i^l}j_l(k(\eta-\eta_0)) \quad (47)$$

leads to

$$\Theta_{lk}(\eta_0) = \int_{\eta_{in}}^{\eta_0} \mathrm{e}^{-\tau(\eta)}\Bigg(\left(-\tau'\frac{1}{4}\delta_{rk} - A'_k + \frac{F'_k k^2}{6}\right)j_l(k(\eta-\eta_0))$$
$$-\left(\tau' v_{rk} + A_k\right)k j'_l(k(\eta-\eta_0)) + \frac{F'_k}{2}k^2 j''_l(k(\eta-\eta_0))\Bigg)d\eta. \quad (48)$$

One may rewrite Eq. (48) in the terms of invariant potentials, densities and velocities (29):

$$\Theta_{lk}(\eta_0) = \int_{\eta_{in}}^{\eta_0} \mathrm{e}^{-\tau(\eta)}\Bigg(-\tau'(\frac{\tilde{\delta}_{rk}}{4} + \Phi_k)j_l(k(\eta-\eta_0))$$
$$-\tau'\tilde{v}_{rk}k\, j'_l(k(\eta-\eta_0)) + (\Phi'_k + \Psi'_k)j_l(k(\eta-\eta_0))\Bigg)d\eta. \quad (49)$$

The integrand expressions in (48) and (50) differ by a total derivative, which does not contribute to the integral because $\mathrm{e}^{-\tau(\eta_{in})} \approx 0$ at the lower limit, and the Bessel function $j_l(0) = 0$ for $l > 0$ at the upper limit.

According to (30), the invariant potentials Ψ and Φ equal zero in our simplified consideration when only $\delta_{rk}(\eta_{in})$ is nonzero. Thus, there is no the Sachs-Wolf effect [2] and the expression (50) is reducible to

$$\Theta_{lk}(\eta_0) = \int_0^{\eta_0} (-\tau')e^{-\tau(\eta)}\left(\frac{\tilde{\delta}_{rk}}{4} j_l(k(\eta-\eta_0)) + \tilde{v}_{rk}k\, j_l'(k(\eta-\eta_0))\right)d\eta$$
$$\approx \frac{\tilde{\delta}_{rk}(\eta_r)}{4} j_l(k(\eta_r-\eta_0)) + \tilde{v}_{rk}(\eta_r)k\, j_l'(k(\eta_r-\eta_0)), \quad (50)$$

where the fact is used that the visibility function $g(\eta) = -\tau' e^{-\tau(\eta)}$[6] is peaked near last scattering surface η_r. On the other hand, the integral $\int g(\eta)d\eta = 1$, and thereby, it is like the Dirac delta-function $g(\eta) = \delta(\eta - \eta_r)$.

Using the expressions for $\tilde{\delta}_{rk}$ and $\tilde{v}_{rk}$ from (30), we obtain the expressions for the coefficients

$$C_l = \frac{2}{\pi}\int_0^{\infty} < \Theta_{lk}(\eta_0) >^2 k^2 dk$$
$$= \frac{2}{\pi}\int_0^{\infty}\left|\frac{(12\mathcal{H}^2+k^2)\cos\left(\frac{k(\eta_r-\eta_{in})}{\sqrt{3}}\right) + 3\sqrt{3}\mathcal{H}k\sin\left(\frac{k(\eta_r-\eta_{in})}{\sqrt{3}}\right)}{4\left(3\mathcal{H}^2+k^2\right)} j_l(k(\eta_r-\eta_0))\right.$$
$$\left. + \frac{9\mathcal{H}k\cos\left(\frac{k(\eta_r-\eta_{in})}{\sqrt{3}}\right) - \sqrt{3}\left(12\mathcal{H}^2+k^2\right)\sin\left(\frac{k(\eta_r-\eta_{in})}{\sqrt{3}}\right)}{4\left(3\mathcal{H}^2+k^2\right)} j_l'(k(\eta_r-\eta_0))\right|^2$$
$$\mathcal{P}(k,\eta_{in})\frac{dk}{k}, \quad (51)$$

where $\mathcal{P}(k, \eta_{in}) = k^3\langle\delta_{rk}(\eta_{in})\delta^*_{rk}(\eta_{in})\rangle$ is a primordial fluid spectrum which serves as an initial condition for the plasma perturbations considered in the previous section.

3.1 *Effect of the Finite Thickness of the Last Scattering Surface*

A real-world visibility function $g(\eta)$ is not exactly the Dirac delta-function, but it is smeared over a finite region of η. One may approximately assume that it has the Gaussian form

$$g(\eta) = -\tau'(\eta)\exp(-\tau) = \frac{1}{\Delta\eta_r\sqrt{2\pi}}\exp\left(-\frac{(\eta-\eta_r)^2}{2\Delta\eta_r^2}\right), \quad (52)$$

[6]The visibility function gives the probability of a CMB photon scattering out of the line of sight within of a $d\eta$−layer on the last scattering surface [1].

where $\Delta\eta_r$ is a width of the last scattering surface. That corresponds to

$$\tau(\eta) = -\ln\left(\frac{1}{2} + \frac{1}{2}\mathrm{erf}\left(\frac{\eta - \eta_r}{\sqrt{2}\,\Delta\eta_r}\right)\right). \tag{53}$$

Let us consider the exact integral

$$\int_{-\infty}^{\infty} g(\eta)\mathrm{e}^{ik(\eta-\eta^*)}d\eta = \exp\left(-k^2\Delta\eta_r^2/2\right)\mathrm{e}^{ik(\eta_r-\eta^*)}. \tag{54}$$

As it is seen (Eq. 54), the variable η is changed by η_r in the expression $\mathrm{e}^{ik(\eta-\eta^*)}$ after integration, and besides a suppression factor appears.

The expression (50) contains the exponents $\mathrm{e}^{i(k\pm k/\sqrt{3})\eta}$ originating from both Bessel functions and $\tilde{\delta}_{\boldsymbol{k}}$. Thus, the suppression factor $\mathrm{e}^{-(k\pm k/\sqrt{3})^2\Delta\eta_r^2/2}$ appears in (50) as a result of integration, which has to be introduced into the integrand of (51). The overall damping factor originates from both Silk dumping and finite width of the last scattering surface, but the last gives the main contribution. The calculation of the last scattering surface width has to take into account the process of hydrogen recombination. In the standard ΛCDM model, one needs using the kinetic equations involving at least three levels of the hydrogen atom. The Milne-like universe expands at $\sqrt{z_r}$−times slower than the standard ΛCDM one. Thus, the Saha equilibrium equation [1, 3]

$$\frac{n_p n_e}{n_H} = \frac{X_e^2}{1-X_e}n_b = \left(\frac{T m_e}{2\pi}\right)^{3/2}\exp\left(-\frac{B_H}{T}\right) \tag{55}$$

is a good estimation, where n_p is a proton density, and n_H is a density of neutral atoms.

Equation (55) allows obtaining the hydrogen ionization degree $X_e = n_p/n_b$, where $n_b = n_p + n_H$. An optical depth [1] is calculated as

$$\tau(\eta) = \sigma_T \int_{\eta}^{\eta_0} n_b(\eta')X_e(\eta')a(\eta')d\eta', \tag{56}$$

where n_b scales as $n_b(\eta) = n_{b0}/a^3(\eta)$ and n_{b0} is given by (32). The visibility function for different values of the matter density is shown in Fig. 1. As one can see the width $\Delta\eta_r$ of the Gaussian approximation is 0.05 for $\Omega_b = 0.03$ and 0.03 for $\Omega_b = 0.3$. In the last case, the visibility function has non-Gaussian shape. However, the initial stage of recombination affects mainly the "left front" of the visibility function which becomes "sharper" and can be approximated by a Gaussian function shown in Fig. 1b.

The expression (54) is exact only for averaging of the exponent, however it is approximately valid and for more complicated expressions like the integrand of (50). As one can see from Fig. 2, the lowest suppression factor $\mathrm{e}^{-(k-k/\sqrt{3})^2\Delta\eta_r^2/2}$ should be taken for the calculations.

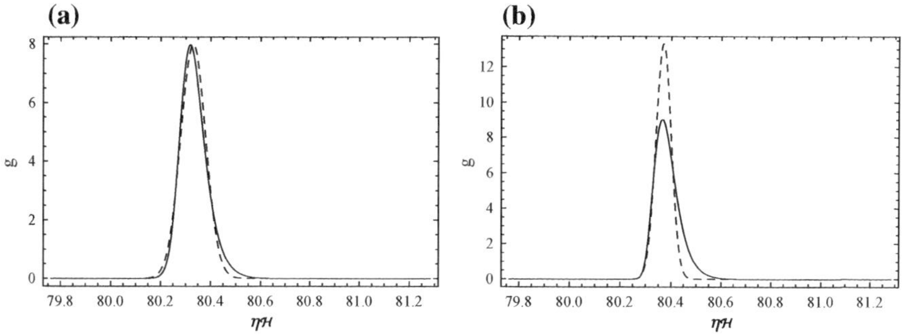

Fig. 1 Visibility function $g(\eta) = -\tau'(\eta)\exp(-\tau)$ for different baryon density (32) **a** $\Omega_b = 0.03$, **b** $\Omega_b = 0.3$ (solid curves). Dashed curves are Gaussian approximations (52) with **a** $\Delta\eta_r = 0.05$ and **b** $\Delta\eta_r = 0.03$

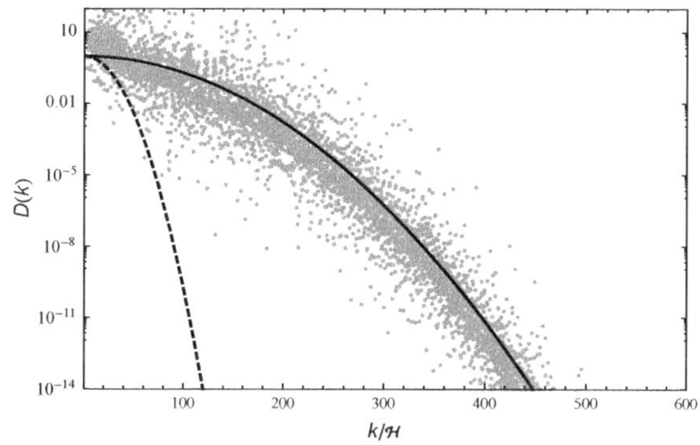

Fig. 2 Calculated damping factor due to finite width of the last scattering surface $D(k) = \left(\int_0^{\eta_0} g(\eta)\left(\frac{\tilde{\delta}_{r\mathbf{k}}(\eta)}{4} j_l(k(\eta-\eta_0)) + \tilde{v}_{r\mathbf{k}}(\eta) k\, j_l'(k(\eta-\eta_0))\right)d\eta\right)^2 / \left(\frac{\tilde{\delta}_{r\mathbf{k}}(\eta_r)}{4} j_l(k(\eta_r - \eta_0)) + \tilde{v}_{r\mathbf{k}}(\eta_r) k\, j_l'(k(\eta_r - \eta_0))\right)^2$ for $\Delta\eta_r = 0.03$, $l = 300$. Dashed and solid curves correspond to $D(k) = \exp\left(-(k + k/\sqrt{3})^2 \Delta\eta_r^2\right)$ and $D(k) = \exp\left(-(k - k/\sqrt{3})^2 \Delta\eta_r^2\right)$ respectively

4 Results and Discussion

A distance from the last scattering surface to the present time observer is $\eta_0 - \eta_r$. For the Milne-like universe (12) these distances are $\eta_0 = \frac{1}{\mathscr{H}} \ln \frac{1}{B}$ and $\eta_r = \frac{1}{\mathscr{H}} \ln \frac{a_r}{B}$ respectively. Thus, one has $\eta_0 - \eta_r \sim \mathscr{H}^{-1} \ln z_r \sim 7\mathscr{H}^{-1}$ independent of B.

To calculate the spectrum according to (51), one needs knowing the initial spectrum. The standard model of cosmological inflation gives almost flat spectrum, i.e., $\mathscr{P}(k) \approx const$ and the oscillations in the observed CMB anisotropy spectrum are interpreted as a result of acoustic oscillation of the photon-baryon plasma. There is a principled difference between the standard model and the linear cosmology consid-

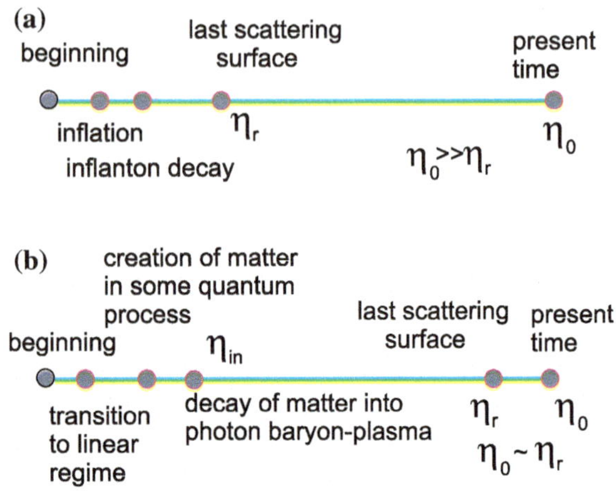

Fig. 3 Schematic representation of the time scales in the **a** standard ΛCDM and **b** linear cosmologies respectively

ered here. In the standard model, the typical angular scale is $\theta \sim \frac{\eta_r - \eta_{in}}{\eta_0 - \eta_r} \sim \frac{\eta_r}{\eta_0}$. As a consequence of $\eta_r << \eta_0$ in the ΛCDM model, one may obtain the angular scale of $\theta \sim 1^\circ$ coinciding with the experimental one. In the linear cosmology $\eta_r \sim \eta_0$ (see scheme in Fig. 3) and the spectrum oscillations should have another origin. In particular, they could originate from the oscillations of the initial spectrum $\mathscr{P}(k, \eta_{in})$, which can be taken in the form

$$\mathscr{P}(k, \eta_{in}) = 3 \times 10^{-7} |\sin k\eta_{in}|^2. \tag{57}$$

For the dependence (12), one has to take $\eta_{in} \sim 0.06$ to obtain experimentally observed angular scale, that gives $\theta \sim \frac{\eta_{in}}{\eta_0 - \eta_r} \sim 0.4^\circ$.

It is easy to calculate cosmic (i.e. physical) time t_{in} corresponding to the conformal time η_{in}. Integrating with (12) gives $t_{in} = \int_0^{\eta_{in}} a(\eta) d\eta \approx B\, \eta_{in}$, where it is taken into account that $\mathscr{H}\eta_{in} \ll 1$. For instance, taking $\eta_{in} = 0.06/\mathscr{H}$ and $B = 3.8 \times 10^{-38}$ gives $t_{in} = 5.9 \times 10^{-22}$ s, which corresponds to the lifetime of the Higgs boson $t_H = 2\pi/\Gamma_H$, where $\Gamma_H = 7$ MeV. Here, it is implied that Higgs bosons are created initially [28], then decay into another particles and, finally, into the baryons and photons. Taking another value of B requires connecting η_{in} with another physical process.

The initial spectrum (57) has to be multiplied by the damping factor[7]

$$D(k) \approx \exp\left(-(k - k/\sqrt{3})^2 \Delta\eta_r^2\right) \approx \exp\left(-k^2/80\right). \tag{58}$$

and substituted into Eq. (51). We do not predict absolute values, and the coefficient in (57) is taken to reproduce only highest first CMB peak. The result, shown in Fig. 4a demonstrates a too strong suppression of higher harmonics in comparison with the observational data. To improve the agreement, one may take a rising initial spectrum

[7]The case of the best agreement with the observational data is considered: $\Omega_m = 0.3$, $\Delta\eta_r = 0.03$.

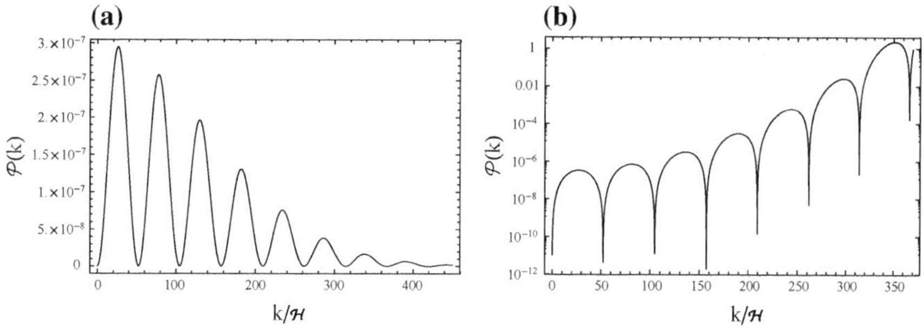

Fig. 4 **a** Initial spectrum multiplied by the all damping factors, i.e., the resulting spectrum $\mathscr{P}(k) = 3 \times 10^{-7} |\sin 0.06k|^2 \exp(-k^2/200^2)$, which reproduces the observational data qualitatively. **b** Rising initial spectrum $\mathscr{P}(k) = 3 \times 10^{-7} |\sin 0.06k|^2 \exp(k^2/87^2)$. It is seen, that the perturbations with $k > 350\mathscr{H}$ lie in the nonlinear region, because $\mathscr{P}(k) > 1$

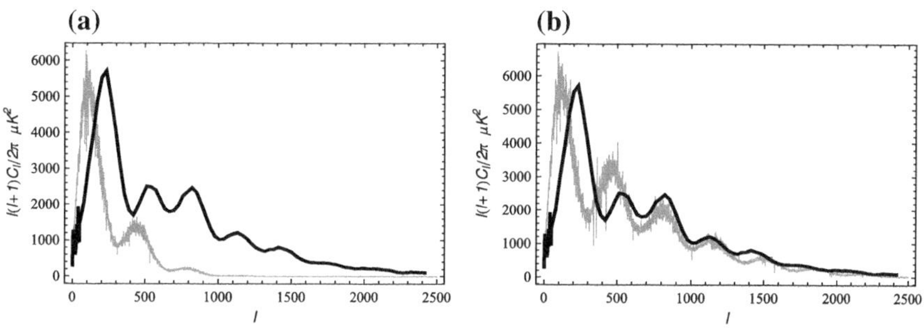

Fig. 5 Cosmic microwave anisotropy spectrum calculated within the framework of the linear Milne-like cosmology (gray noisy curve). Black curve corresponds to the "Planck"-satellite data [30]. The quantities C_l are dimensionless (the multiplication by the squared present CMB temperature gives the dimensional C_l). **a** Corresponds to the initial spectrum (57), **b** corresponds to the rising spectrum (59)

$$\mathscr{P}(k, \eta_{in}) = 3 \times 10^{-7} |\sin k\eta_{in}|^2 \exp\left(k^2/\kappa_{in}^2\right). \tag{59}$$

with $\kappa_{in} = 87$ in order to obtain the overall damping factor about of $\exp\left(-k^2/200^2\right)$, because $80^{-2} - 87^{-2} \approx 200^{-2}$ (Fig. 4b).

The results of calculation with this formula is shown in Fig. 5. The *Planck*-satellite data give a very precise measurement of the CMB anisotropy [2, 29–31]. One can see the qualitative coincidence with the spectrum observed by the *Planck*-satellite. The positions of the peaks are shifted relatively observed ones. However, it is no wonder because the model considered is rough and requires further development. At least, the model needs taking into account the baryonic content explicitly. Of course, no analytic solutions for perturbations could be found with this complication. The Silk dumping and the finite width of the last scattering surface have to be taken into account more accurately. Besides, more complicated models of the initial spectrum have to be considered (Fig. 4).

From a fundamental point of view, it could imagine some breathtaking physics like the inflation theory. However, it could be quite different, because the inflation cannot produce, a "violet", i.e., rising with k, initial spectrum (59). In principle, the linear cosmology needs no inflation, because the scales of perturbations modes always remain within the horizon and there is no need in any model like inflation for the superhorizon spectral modes. Thus, the liner universe seems in some sense simpler compared to the standard ΛCDM model. However, the most fundamental problem of the linear cosmology is a requirement of more accurate consideration of vacuum perturbations with taking into account the quantum properties of the vacuum. The above simple model of vacuum as a fluid with the equation of state $w = -1/3$ is an only very rough heuristic approximation.

Unfortunately, well-known software packages such as CAMB [32] and CMB-FAST [33] are absolutely useless for the calculation of CMB spectrum in the linear cosmology because they assume a quite different formation mechanism for the CMB spectrum peaks. It seems that the tools for the ionization history analysis, such as RECFAST [34], also have to be modified to take into account more than three levels of the hydrogen atom. It results from the fact that partially ionized hydrogen plasma is closer to thermal equilibrium due to the slower expansion of the Milne-like universe and, thereby, more hydrogen levels are populated. It seems that the pure equilibrium Saha formula used above gives a sufficiently good approximation in this case.

It should also to do some notes about distortion of the CMB spectrum from blackbody one [35]. The expected distortion of the spectrum caused by hydrogen recombination should be mach smaller than that in the ΛCDM model.

Appendix

The expression for the perturbation of the vacuum density is given by

$$\delta\rho_{vk} = \frac{B^2\mathscr{H}^3{M_p}^2 e^{-4\eta\mathscr{H}}}{8{\rho_{r0}}^2\left(3\mathscr{H}^2+k^2\right)}\Big(-C_1\big(B^2\mathscr{H}{M_p}^2e^{2\eta\mathscr{H}}\left(3\mathscr{H}^2+2i\sqrt{3}\mathscr{H}k-k^2\right)$$
$$-6\mathscr{H}\rho_{r0}+2i\sqrt{3}k\rho_{r0}\big)e^{-i\frac{\eta k}{\sqrt{3}}}+C_2\big(-B^2\mathscr{H}{M_p}^2e^{2\eta\mathscr{H}}\left(3\mathscr{H}^2-2i\sqrt{3}\mathscr{H}k-k^2\right)$$
$$+6\mathscr{H}\rho_{r0}+2i\sqrt{3}k\rho_{r0}\big)e^{i\frac{\eta k}{\sqrt{3}}}\Big)+\frac{B^2\mathscr{H}^3{M_p}^2e^{-2\eta\mathscr{H}}}{4\rho_{r0}\left(3\mathscr{H}^2+k^2\right)}\Big(C_3(3\mathscr{H}+i\sqrt{3}k)e^{-i\frac{\eta k}{\sqrt{3}}}$$
$$+C_4(3\mathscr{H}-i\sqrt{3}k)e^{i\frac{\eta k}{\sqrt{3}}}\Big)-\frac{k^4{M_p}^2e^{-3\eta\mathscr{H}}}{18B^2}\left(Z_1e^{-\eta\sqrt{\mathscr{H}^2+\frac{k^2}{3}}}+Z_2e^{\eta\sqrt{\mathscr{H}^2+\frac{k^2}{3}}}\right),$$

Then

$$V_{vk} = -\frac{B^2\mathscr{H}^2 M_p^2 e^{-4\eta\mathscr{H}}}{24\rho_{r0}^2\left(3\mathscr{H}^2 k + k^3\right)}\left(C_1\left(B^2\mathscr{H} k M_p{}^2 e^{2\eta\mathscr{H}}\left(3\mathscr{H}^2 + 2i\sqrt{3}\mathscr{H}k - k^2\right)\right.\right.$$
$$\left. - 2i\rho_{r0}(6\sqrt{3}\mathscr{H}^2 - 3i\mathscr{H}k + \sqrt{3}k^2)\right)e^{-i\frac{\eta k}{\sqrt{3}}} + C_2\left(B^2\mathscr{H}kM_p{}^2 e^{2\eta\mathscr{H}}\left(3\mathscr{H}^2\right.\right.$$
$$\left.\left.\left. -2i\sqrt{3}\mathscr{H}k - k^2\right) + 2i\rho_{r0}(6\sqrt{3}\mathscr{H}^2 + 3i\mathscr{H}k + \sqrt{3}k^2)\right)e^{i\frac{\eta k}{\sqrt{3}}}\right)$$
$$+\frac{B^2\mathscr{H}^2 M_p{}^2 e^{-2\eta\mathscr{H}}}{12\rho_{r0}\left(3\mathscr{H}^2 + k^2\right)}\left(C_3(3\mathscr{H} + i\sqrt{3}k)e^{-i\frac{\eta k}{\sqrt{3}}} + C_4(3\mathscr{H} - i\sqrt{3}k)e^{i\frac{\eta k}{\sqrt{3}}}\right)$$
$$+\frac{k^2 M_p{}^2 e^{-3\eta\mathscr{H}}}{54B^2}\left(Z_1\left(\sqrt{3}\sqrt{3\mathscr{H}^2 + k^2} + 3\mathscr{H}\right)e^{-\eta\sqrt{\mathscr{H}^2 + \frac{k^2}{3}}}\right.$$
$$\left.+ Z_2\left(\sqrt{3}\sqrt{3\mathscr{H}^2 + k^2} - 3\mathscr{H}\right)e^{\eta\sqrt{\mathscr{H}^2 + \frac{k^2}{3}}}\right),$$

$$F_k = \frac{B^4 e^{-2\eta\mathscr{H}}}{4\rho_{r0}{}^2\left(3\mathscr{H}^2 k + k^3\right)}\left(C_1\left(B^2\mathscr{H}M_p{}^2 e^{2\eta\mathscr{H}}\left(-3i\sqrt{3}\mathscr{H}^2 + 6\mathscr{H}k\right.\right.\right.$$
$$\left.\left. + i\sqrt{3}k^2\right) - 6i\sqrt{3}\mathscr{H}\rho_{r0} - 6k\rho_{r0}\right)e^{-i\frac{\eta k}{\sqrt{3}}} + C_2\left(B^2\mathscr{H}M_p{}^2 e^{2\eta\mathscr{H}}\left(3i\sqrt{3}\mathscr{H}^2\right.\right.$$
$$\left.\left.\left. + 6\mathscr{H}k - i\sqrt{3}k^2\right) + 6i\sqrt{3}\mathscr{H}\rho_{r0} - 6k\rho_{r0}\right)e^{i\frac{\eta k}{\sqrt{3}}}\right) - \frac{3B^4}{2k\rho_{r0}\left(3\mathscr{H}^2 + k^2\right)}$$
$$\left(C_3(k - i\sqrt{3}\mathscr{H})e^{-i\frac{\eta k}{\sqrt{3}}} + C_4(k + i\sqrt{3}\mathscr{H})e^{i\frac{\eta k}{\sqrt{3}}}\right) + e^{-\mathscr{H}\eta}\left(Z_1 e^{-\frac{\eta\sqrt{3\mathscr{H}^2 + k^2}}{\sqrt{3}}}\right.$$
$$\left.+ Z_2 e^{\frac{\eta\sqrt{3\mathscr{H}^2 + k^2}}{\sqrt{3}}}\right),$$

$$A_k = \frac{e^{-2\eta\mathscr{H}}}{24\rho_{r0}^2}\left(C_1\left(-6B^4\rho_{r0} + B^6\mathscr{H}M_p{}^2 e^{2\eta\mathscr{H}}\left(3\mathscr{H} + i\sqrt{3}k\right)\right)e^{-i\frac{\eta k}{\sqrt{3}}}\right.$$
$$\left. + C_2\left(-6B^4\rho_{r0} + B^6\mathscr{H}M_p{}^2 e^{2\eta\mathscr{H}}\left(3\mathscr{H} - i\sqrt{3}k\right)\right)e^{i\frac{\eta k}{\sqrt{3}}}\right)$$
$$-\frac{B^4}{4\rho_{r0}}\left(C_3 e^{-\frac{i\eta k}{\sqrt{3}}} + C_4 e^{\frac{i\eta k}{\sqrt{3}}}\right).$$

References

1. V. Mukhanov, *Physical Phoundations of Cosmology* (Cambridge University Press, Cambridge, 2005)
2. R. Durrer, *The Cosmic Microwave Background* (Cambridge University Press, Cambridge, 2008)
3. S. Dodelson, *Modern Cosmology* (Elsevier, Amsterdam, 2003)
4. E.A. Milne, *Kinematic Relativity* (The Clarendon Press, Oxford, 1935)
5. E.A. Milne, *Relativity, Gravitation and World-Structure* (The Clarendon Press, Oxford, 1935)

6. A. Dev, M. Safonova, D. Jain, D. Lohiya, Cosmological tests for a linear coasting cosmology. Phys. Lett. B **548**, 12–18 (2002)
7. P. Singh, D. Lohiya, Constraints on Lepton asymmetry from nucleosynthesis in a linearly coasting cosmology. J. Cosmol. Astropart. Phys. **05**, 061 (2015)
8. A. Benoit-Lévy, G. Chardin, The Dirac-Milne cosmology. Int. J. Mod. Phys.: Conf. Ser. **30**, 1460272 (2014)
9. F. Melia, On recent claims concerning the $R_h = ct$ Universe. Monthly Not. R. Astron. Soc. **446**, 1191–1194 (2015)
10. F. Melia, The linear growth of structure in the $R_h = ct$ universe. Monthly Not. R. Astron. Soc. **464**, 1966–1976 (2017)
11. D.L. Shafer, Robust model comparison disfavors power law cosmology. Phys. Rev. D **91**, 103516 (2015)
12. G.R. Bengochea, G. Leon, Puzzling initial conditions in the $R_h = ct$ model. Eur. Phys. J. C **76**, 626 (2016)
13. I. Tutusaus, B. Lamine, A. Blanchard, A. Dupays, Y. Zolnierowski, J. Cohen-Tanugi, A. Ealet, S. Escoffier, O. Le Fèvre, S. Ilić, A. Pisani, S. Plaszczynski, Z. Sakr, V. Salvatelli, Th Schücker, A. Tilquin, J.-M. Virey, Power law cosmology model comparison with CMB scale information. Phys. Rev. D **94**, 103511 (2016)
14. M.V. John, Realistic coasting cosmology from the Milne model. arXiv:1610.09885 [astro-ph.CO]
15. E. Ling, Milne-like spacetimes and their role in Cosmology. arXiv:1706.01408 [gr-qc]
16. S.L. Cherkas, V.L. Kalashnikov, Universe driven by the vacuum of scalar field: VFD model, in Proceedings of International Conference "Problems of Practical Cosmology", St.-Petersburg, 2008, vol. II (Russian Geographical Society, Saint Petersburg, 2008), p. 135, arXiv: astro-ph/0611795
17. F. Melia, The cosmic equation of state. Astrophys. Space Sci. **356**, 393–398 (2015)
18. S.L. Cherkas, V.L. Kalashnikov, Theory of gravity admitting arbitrary choice of the energy density level. arXiv:1609.00811 [gr-qc]
19. S. Weinberg, The cosmological constant problem. Rev. Mod. Phys. **61**, 1–23 (1989)
20. N.D. Birrell, P.C.W. Davis, *Quantum Fields in Curved Space* (Cambridge University Press, Cambridge, 1982)
21. Y.B. Zel'dovich, The cosmological constant and the theory of elementary particles. Sov. Phys. Usp. **11**, 381–393 (1968)
22. R.J. Adler, B. Casey, O.C. Jacob, Vacuum catastrophe: an elementary exposition of the cosmological constant problem. Am. J. Phys. **63**, 620–626 (1995)
23. L.D. Landau, E.M. Lifshitz, *The Classical Theory of Fields* (Butterworth-Heinemann, Oxford, 2000)
24. D. Lehmkhul, G. Schiemann, E. Scholz (eds.), *Towards a Theory of Spacetime Theories* (Springer, Boston, 2010)
25. E. Anderson, *The Problem of Time* (Springer, Swizerland, 2017)
26. H.D. Zeh, *The Physical Basis of the Direction of Time* (Springer, Berli, 2007)
27. R. Arnowitt, S. Deser, C.W. Misner, The dynamics of general relativity, in *Gravitation: An Introduction to Current Research*, ed. by L. Witten (Wiley, New York, 1962), chap. 7, p. 227. arXiv:gr-qc/0405109
28. S.L. Cherkas, V.L. Kalashnikov, Matter creation and primordial CMB spectrum in the inflationless Milne-like cosmologies. Proc. Nat. Acad. Sci. Belarus (Phys. Math. Ser.) **4**, 88–97 (2017). https://arxiv.org/abs/1707.06073
29. ESA: Planck collaboration results. https://www.cosmos.esa.int/web/planck/publications
30. N. Aghanim et al., Planck 2015 results. XI. CMB power spectra, likelihoods, and robustness of parameters. A&A **A11** 594 (2016)
31. W. Hu, N. Sugiyama, Small scale cosmological perturbations: an analytic approach. Astrophys. J. **471**, 542–570 (1996)
32. A. Lewis, A. Challinor, A. Lasenby, Efficient computation of CMB anisotropies in closed FRW models. Astrophys. J. **538**, 473–476 (2000)

33. U. Seljak, M. Zaldarriaga, A line of sight approach to cosmic microwave background anisotropies. Astrophys. J. **469**, 437–444 (1996)
34. S. Seager, D.D. Sasselov, D. Scott, A new calculation of the recombination epoch. Astrophys. J. **523**, L1–L5 (1999)
35. J.A. Rubino-Martin, J. Chluba, R.A. Sunyaev, Lines in the cosmic microwave background spectrum from the epoch of cosmological hydrogen recombination. Mon. Not. R. Astron. Soc. **371**, 1939–1952 (2006)

Printed by Printforce, the Netherlands